Spectral Expansion of the Transfer Matrices of Gibbs Fields

R.A. Minlos

Cambridge Scientific Publishers

Reviews in Mathematics and Mathematical Physics
Volume 16, Part 2

First published in Soviet Scientific Reviews: Section C © 1988 OPA (Overseas Publishers Association) N.V. Published by license under the Harwood Academic Publishers imprint.

Printed in UK

ISBN 978-1-904868-99-6 Paperback

Cambridge Scientific Publishers Ltd
45 Margett Street
Cottenham
Cambridge CB24 8QY
UK
www.cambridgescientificpublishers.com
janie.wardle@cambridgescientificpublishers.com

Introduction to the series

In recent years, many young excellent mathematicians both in Russia and abroad made an excellent name for themselves in the international mathematical community. *Reviews in Mathematics and Mathematical Physics* will publish the most outstanding and recent results. Not only new research, but also a detailed background of a problem will be presented to give an insight into a particular direction. This would enable a better grasp of the state of the art in a particular field and an easier positioning of the results discussed.

To maintain high scientific standards, all works published in *Reviews in Mathematics and Mathematical Physics* are peer reviewed by two or three internationally recognized mathematicians. Many of the works presented by Russian mathematicians have been discussed at the seminars at the Faculty of Mechanics and Mathematics, Moscow State University and Steklov Mathematical Institute, Russian Academy of Sciences.

Reviews in Mathematics and Mathematical Physics also plan publications of review papers by oustanding mathematicians to ensure a broader view of the development of modern mathematics and mathematical physics.

A.T. Fomenko
Editor

Contents

Carlo Boldrighini (Istituto Nazionale di Alta Matematica, INdAM, Roma, Italy), Vadim A.Malyshev (Lomonosov Moscow State University, Russia), Alessandro Pellegrinotti (Universita Roma Tre, Italy), Suren K. Poghosyan (Institute of Mathematics of the NAS RA, Yerevan, Armenia), Yakov G. Sinai (Princeton University, USA), Valentin A. Zagrebnov (Institut de Mathématiques de Marseille, France), Elena A. Zhizhina (Institute for Information Transmission Problems, Moscow, Russia)

"Robert Adol'fovich Minlos (1931–2018) – His Work and Legacy"

EMS Newsletter, June 2018, pp. 22–27
Reprinted with permission and with our sincere thanks to Editor and Editorial Board of European Mathematical Society Newsletter

Reviews in Maths & Maths Phys.
2019, Vol 16, Part 2, pp. 1–68
Reprints available directly from publisher
Photocopying permitted by license only

Preface

This edition is a tribute to Professor Robert A. Minlos (1931–2018) in recognition of his achievements and significant contribution to mathematics, research and education.

A review of the life and work of Professor Robert A. Minlos is included in this issue after the classic paper "Spectral Expansion of the Transfer Matrices of Gibbs Fields". This article, in honor of Professor Robert A. Minlos, and entitled: Robert Adol'fovich Minlos (1931–2018) – His Work and Legacy; was originally published in the June 2018 issue of the European Mathematical Society Newsletter and we would like to convey our sincere thanks to the editor, Professor Valentin Zagrebnov and the European Mathematical Society and EMS Publisher for permission to reprint this article in the series Reviews in Mathematics and Mathematical Physics.

Spectral expansion of the transfer-matrix of Gibbs fields – R.A. Minlos was first published in *Mathematical Physics Reviews (Soviet Scientific Reviews)* in 1988 and presented a review of a series of papers studying the structure of the spectrum of transfer matrices (stochastic operators) of lattice Gibbs fields. The investigations were initiated by R.A. Minlos and I.G. Sinai, authors of the paper: *Study of the spectra of stochastic operators arising in lattice models of a gas* which examined the leading (one and two particle) invariant subspaces of the transfer matrix of the Gibbs two-dimensional spin field at high temperatures. The basic ideas and methods were developed in the context of more complex lattice models in a range of papers published subsequently, (see references 2–34) and the applications are wide ranging.

This edition has been updated with a supplementary review of recent investigations. The general methods of studying spectral structure of transfer-matrices of Gibbsian fields and related topics have some applications and development in recent papers of the author, his collaborators and students. The following themes and associated references are reviewed:

- The transfer-matrices of Gibbsian fields
- Random walks in random environment
- The asymptotics of decay of correlations for Gibbsian fields at high temperatures
- Inhomogeneous random walk on the lattice
- Some models from solid-state theory
- Stochastic dynamics

Professor Robert A. Minlos was Head of the Dobrushin Mathematical Laboratory of the Institute for Information Transmission Problems, Russian Academy of Sciences and Professor at the Physics and Mathematics Department of the Independent University of Moscow and also taught at Moscow State University (as the professor of Theory of Probability Chair). His scientific areas of research concerned mathematical physics, functional analysis and probability theory. His significant and fundamental results pertained to the theory of generalized random fields, the theory of phase transitions, the investigation of spectrum of transfer-matrix of Gibbsian random fields in statistical physics and quantum field theory and also some quantum-mechanical systems, the random walks in random environment

Professor Robert A. Minlos was an internationally well-known specialist in functional analysis, probability theory and contemporary mathematical physics and an outstanding researcher and teacher. He was also a well-known and established author and editor of many scientific papers, monographs and textbooks including:

- Representation of rotations group and the Lorentz group – I.M. Gelfand, R.A. Minlos, Z.J. Shapiro, (FIZMATGIZ 1959 and the English translation published by Pergamon Press in 1963).
- Gibbs Random Fields – V.A. Malyshev and R.A. Minlos (published by Nauka in1985 and the English translation published by Kluwer in 1991).
- Linear operators in infinite-particle systems – V.A. Malyshev and R.A. Minlos; AMS, 1995.
- Introduction to mathematical statistical physics – R.A. Minlos, AMS, 2000.
- Study of the spectra of stochastic operators arising in lattice models of a gas – R.A. Minlos and Ya. G. Sinai; Teor. Mat. Fizika 2, 230–243.

SPECTRAL EXPANSION OF THE TRANSFER MATRICES OF GIBBS FIELDS

R.A. MINLOS

Department of Mathematics and Mechanics, Moscow State University, Moscow

Abstract

Investigations of the structure of the spectrum of transfer matrices (stochastic operators) of lattice Gibbs fields are surveyed. Cluster expansion of the transfer matrix, invariant cluster R-particle subspaces of the transfer matrix, and cluster operators in the "p-representation" are considered. Some supplementary topics (e.g. the Bethe-Salpeter kernel and other examples of cluster operators) are reviewed.

Contents

1. Introduction

This paper surveys the studies of a small group of Moscow mathematicians into the structure of the spectrum of transfer matrices (stochastic operators) of lattice Gibbs fields. The studies were initiated by the paper by Minlos and Sinai [1], which examined the leading (one- and two-particle) invariant subspaces of the transfer matrix of the Gibbs two-dimensional spin field at high temperatures. The basic ideas and methods of [1] were developed in the context of more complex lattice models in a wide range of papers [2]–[34]. The study of the spectrum of transfer matrices in these papers uses the technique of cluster expansions of Gibbs fields (see the book by Malyshev and Minlos [35]), and also devices of spectral analysis of finite-particle systems, mainly developed in the context of studies of the Schrödinger operator (see Reed and Simon, [43]).

The interest in this range of topics is linked with the so-called "Euclidean Markov approach" to models of quantum field theory and statistical physics. In accordance with this strategy, for certain physical "infinite-particle" systems (specified in continuous space R^ν or on the lattice Z^ν) it is possible to construct a Markov random field in space-time $R^\nu \times R^1$ (or $Z^\nu \times R^1$) in such a way that the generator of the group of its transition operators T^t (the transfer-matrix group) is unitarily equivalent to the (renormalized) Hamiltonian of the system in the "ground state" (i.e., the operator describing the lowest excited energy levels of the system).

In this paper we consider fields which are defined in discrete space-time, since the case of continuous space (or continuous time) can be reduced to the discrete case by "coarsening" the set of "spins" (field values).

We shall only take the case of a Gibbs field which is a small disturbance of an independent field (except for the one example in Section 3.3). As applied to the case of Gibbs fields at low temperatures (disturbance of the ground state), or to the case of disturbances of Gaussian fields, our method is still insufficiently developed.

The important general concept of a cluster operator has crystallized during the study of transfer matrices. The class of these operators includes virtually all the operators encountered in mathematical physics. At the same time, it has an inner harmony which certainly merits independent study, apart from any physical context. The basic hypothesis (which has been checked for certain particular examples) consists of the fact that the spectrum of the self-adjoint so-called additive cluster operator (a special class of cluster operators which contains most Hamiltonians, see below) has a "corpuscular" structure: it is determined by its own "one-particle branches" and consists of all possible "tensor sums" of these branches. In other words, the state of the system can be described as a finite set of quasi-particles, the total energy of the set being made up of the energies of the individual quasi-particles. Processes of interaction between quasi-particles in the sets then lead to their scattering (with possible conversion of some quasi-particles into others). This picture of the spectrum of a "channel" scattering is well known for the case of the many-particle Schrödinger operator.

Note that, in many studies of bound states of transfer matrices in quantum models of field $P(s)_2$ (Glimm, Jaffe [36], Spenser [37], Spenser, Zirilli [38], Dimock, Eckmann [39]) a method linked with the concept of Bethe-Salpeter kernels has been used. The same method was later used by Schor [40] and O'Carroll, e.g., [41], to study the bound states of the transfer matrix of discrete Gibbs fields. At the end of our paper we briefly explain the connection between this method and our constructions.

The expressions are separately numbered in each section.

To refer to an expression from another section, we add to the expression number its section number, e.g., (6.3) means expression (6) of Section 3.

2. The Transfer Matrix of a Gibbs Random Field (General Concepts)

2.1. The Transfer Matrix (Stochastic Operator) of a Random Markov Field

Let $\xi = \{\xi_x, x \in Z^{v+1}\}$ be a random field, defined on the cubic $(v+1)$-dimensional lattice Z^{v+1}, with values in a (measurable) space X (the space of "spins"). We agree to call the $(v+1)$-th axis of the lattice Z^{v+1} the time axis, and say that the field is Markov in time, if, given any $t \in Z^1$, and given a fixed configuration $\xi_t = \{\xi_x, x \in Y_t\}$, of the field on the t-th layer of the lattice $Y_t = \{x: x^{(v+1)} = t\} \subset Z^{v+1}$, the values of the field "in the past" and its values "in the future" with respect to Y_t are conditionally independent.

It will henceforth always be assumed that the field ξ is translationally invariant and invariant under "time reversal". This means that transformations in the space $X^{Z^{v+1}} = \Omega$ of all field configurations, which are generated by shifts along Z^{v+1}:

$$(\tau_s \xi)_x = \xi_{x+s}, x, s \in Z^{v+1} \tag{1}$$

and also by "time reversal":

$$(\sigma\xi)_x = \xi_{\theta x} \tag{2}$$

where θ is the reflection of Z^{v+1} in layer Y_0 $(t=0)$, do not change the field probability distribution P.

We denote by $H = L_2(X^{Y_0}, P)$ the Hilbert space of all functionals $f \in L_2$ of the field, dependent only on its restriction $\xi_0 = \xi|_{Y_0}$ onto the zero layer Y_0 of lattice Z^{v+1} (so-called *physical space*) and define the family of operators $\{T_t, t=1, 2, \ldots\}$ in H by the relation

$$T_t f = P_H U_{te_{v+1}} f, f \in H, t=1, 2, \ldots \tag{3}$$

where U_s, $s \in Z^{v+1}$, is the unitary operator in $L_2(\Omega, P)$, generated by shift (1), e_{v+1} is the unit vector along the time axis, and P_H is the projector onto subspace $H \subset L_2(\Omega, P)$. It follows from (3) that the value of the functional $(T_t f)(\xi_0)$ on the configuration of field ξ_0 on the zero layer Y_0 is equal to

$$(T_t f)(\xi_0) = \langle U_{te_{v+1}} f \mid \xi_0 \rangle \tag{4}$$

where $\langle . \mid \xi_0 \rangle$ is the conditional mean over the distribution of the field ξ provided that its restriction $\xi \mid_{Y_0} = \xi_0$ is fixed. It is easily shown that, since the field is Markov and translationally invariant, the operators T_t form a semigroup:

$$T_t = (T_1)^t, \; t = 2, 3, \ldots \tag{5}$$

The operator $T_1 = T$ is in fact called the *transfer matrix* of the field ξ. Its matrix elements $(T f, g)$, $f, g \in H$, are equal to

$$(Tf, g) = \langle (U_{e_{v+1}} f).\bar{g} \rangle \tag{6}$$

($\langle . \rangle$ is the mean over the distribution of the field ξ).

Since the field is translationally invariant, the transfer matrix T commutes with the group of shifts $\{U_s, s \in Y_0 = Z^v\}$ along the layer Y_0:

$$TU_s = U_s T \tag{7}$$

and the field invariance with respect to time reversal (2) implies that the operator T is self-adjoint.

Note. Our definitions can be generalized in a natural way: instead of a Markov random field on the lattice $Z^{v+1} = Z^v \times Z^1$, we can consider a Markov random field $\xi = (\xi(y, t), (y, t) \in E \times Z^1\}$ on the set $E \times Z^1$, where E is a renumberable set. As above, we can speak of the invariance of the field with respect to time reversal, while, if a group G of its automorphisms (most commonly $G = Z^v$) acts in the set E, we can speak of the invariance of the field with respect to the group $G \times Z^1$, acting in $E \times Z^1$. The space H and the transfer matrix T are now introduced in complete analogy with the previous definitions.

2.2. *Gibbs Random Fields*

Let us briefly recall the definition of a Gibbs random field on the lattice Z^{v+1}, in order to agree on our terms and definitions; for details, we refer the reader to the book [35]. Let the space of spins X be furnished with a probability measure v_0: we denote by $\mu_0 = v_0^{Z^{v+1}}$ the measure in space Ω of field configurations, equal to the product of "Z^{v+1}-exemplars" of measure v_0. Further, suppose we are given in the space of configurations Ω the system of functions (potentials)

$$\phi = \{\phi_A, A \subset Z^{v+1}\}, \tag{7^a}$$

marked by certain finite subsets of Z^{v+1} (the potential supports) in such a way that every function ϕ_A depends only on the values of the configuration $\xi \in \Omega$ at points of the set A. For every finite set $\Lambda \subset Z^{v+1}$, we define the "energy of interaction" of configuration ξ "inside Λ"

$$U_\Lambda(\xi) = \sum_{A \leq \Lambda} \phi_A(\xi) \tag{8}$$

and the probability measure μ_Λ in Ω

$$\frac{d\mu_\Lambda}{d\mu_0} = \frac{1}{Z_\Lambda} \exp\{-U_\Lambda(\xi)\} \tag{9}$$

where $Z_\Lambda = \langle e^{-U_\Lambda} \rangle_{\mu_0}$ is a normalizing factor. We call (9) a *finite Gibbs distribution* in Λ, while the weak limit of these distributions

$$\mu = \lim_{\Lambda \nearrow Z^{v+1}} \mu_\Lambda \tag{10}$$

on an unbounded extension of the set Λ is the limiting Gibbs distribution. It can be regarded as the distribution of probability values of the random field ξ on the lattice Z^{v+1} with values in X, which is the so-called *limiting Gibbs field.*

Note. For "small" potentials ϕ, to which most of our constructions refer, a limiting Gibbs field exists. However, if the potential is not "small," then the limit (10) may not exist in the literal sense. In this case, finite Gibbs fields in Λ of a more general type than (9) are considered, by adding to the interaction energy $U_\Lambda(\xi)$ the further interaction

$$\sum_{\substack{A:\, A \cap \Lambda \neq \phi \\ A \bar{\subset} \Lambda}} \phi_A(\xi) \tag{11}$$

of part of the configuration ξ inside Λ with its external part $\bar{\xi} = \xi|_{Z^{v+1} \setminus \Lambda}$, which is assumed to be fixed. The Gibbs distribution $\mu_\Lambda(\cdot \mid \bar{\xi})$ of this field now depends on the external configuration $\bar{\xi} = \bar{\xi}_\Lambda$, and in the passage to the limit $\Lambda \nearrow Z^{v+1}$ it can be chosen in such a way that $\lim_{\Lambda \nearrow Z^{v+1}} \mu_\Lambda(\cdot \mid \bar{\xi}_\Lambda)$ exists. The distributions thus obtained in space Ω (there can be several of them) and the corresponding random fields in $Z^{\nu+1}$ are also called *limiting Gibbs distributions* (*Gibbs fields* or sometimes *pure Gibbs phases*) in the sense of Dobrushin, Lunford, and Rouelle (D.L.R.). In the case of "small" potentials these distributions are all the same and equal to the limit (10).

Henceforth, we shall take the cases when the potential:

1. is finite (i.e., its supports have bounded diameter: diam $A \leq r_0$, where r_0 is a fixed number, the so-called "interaction radius") and every support A is contained entirely in the union of two adjacent layers: $A \subset Y_t \cup Y_{t+1}$.
2. is translationally invariant and invariant under time reversal:

$$\phi_A(\tau_s \xi) = \phi_{A-s}(\xi); \; \phi_A(\sigma\xi) = \phi_{\theta A}(\xi) \tag{12}$$

It is easily seen that a Gibbs random field constructed from such a potential (in the sense of definition (10)), is Markov in time, translationally invariant, and invariant under time reversal.

We shall later be concerned with a special class of potentials (which of course satisfy the general requirements 1) and 2)). Suppose we are given in space X the orthonormalized (with respect to measure ν_0) basis $\{\varphi_\gamma, \gamma \in M\}$, consisting of bounded functions φ_γ, and marked by elements of a denumerable set M. It will be assumed that every element γ is assigned a positive trigger $\ell(\gamma) \geq 0$ (the rank of γ) such that we have

$$\varphi_{\gamma_1} \cdot \varphi_{\gamma_2} = \sum_{\substack{\gamma: \\ |\ell(\gamma_1) - \ell(\gamma_2)| \leq \ell(\gamma) \leq \ell(\gamma_1) + \ell(\gamma_2)}} C^{\gamma}_{\gamma_1,\gamma_2} \varphi_\gamma, \quad \gamma_1, \gamma_2 \in M \tag{13}$$

where the coefficients $C^{\gamma}_{\gamma_1,\gamma_2}$ are bounded in aggregate. With respect to M it will also be assumed that:

1. there exists a marked element $\theta \in M$
   $\ell(\theta) = 0$, $\varphi_\theta \equiv 1$
2. there exists an involution $\gamma \to \gamma^*$ in M, such that $\theta^* = \theta$ and $\ell(\gamma^*) = \ell(\gamma)$, while $\varphi_{\gamma^*} = \bar{\varphi}_\gamma$ (the bar denotes the complex conjugate).
3. the number N_ℓ of elements $\gamma \in M$ of fixed rank ℓ increases as $\ell \to \infty$ not faster than according to a power law:

$$N_e < k\,\ell^s \tag{13a}$$

   where $k > 0$ and $s \geq 0$ are constants.

We consider potentials of the "polynomial" type

$$\phi_A(\xi) = \beta \sum_{\Gamma:\, \mathrm{supp}\, \Gamma = A} B_\Gamma \varphi_\Gamma(\xi) \tag{14}$$

where $\Gamma = \{\gamma(x), x\in Z^{\nu+1}\}$ is a multiindex, i.e., a finite function in $Z^{\nu+1}$ with values in M (the fact that Γ is finite implies that the set supp $\Gamma \equiv \{x: \gamma(x) \neq \theta\}$ is finite), while

$$\varphi_\Gamma = \prod_{x\in Z^{\nu+1}} \varphi_{\gamma(x)}(\xi_x) \tag{15}$$

In (14), B_Γ are coefficients, while β is a parameter (which will henceforth most commonly be assumed small). We know, see e.g., [35], that, when β is sufficiently small, the limiting Gibbs distribution constructed with the help of potential (14) exists.

2.3. Examples

Let us give a few examples of Gibbs fields of the type described.

(i) *Spin models*. Space $X = \{-1, 1\}$, measure $v_0(\{1\}) = v_0(\{-1\}) = 1/2$; basis $\{\varphi_0 \equiv 1, \varphi_1(\sigma) = \sigma; \sigma = \pm 1\}$. The potential is

$$\phi_A(\sigma) = \beta J_A \prod_{x\in A} \sigma_x; \quad \sigma = \{\sigma_x, x\in Z^{\nu+1}\} \tag{16}$$

where J_A are coefficients. When $J_A = 0$ for all A that consist of more than two points, potential (16) is called a pair potential.

(ii) *Model of rotators*. Space $X = T^1$ is a circle, v_0 is the ordinary (normalized) Lebesgue measure in T^1, the basis $\{\varphi_n = e^{in\theta}, \theta\in T^1, n = 0, \pm 1, \pm 2, \ldots\}$. The potential ϕ is the pair potential of "nearest neighbours" (i.e., the ϕ_A differ from zero only for two-point sets $A = \{x_1, x_2\}$, consisting of neighbouring points). Then,

$$\begin{aligned} \phi_{\{x_1,x_2\}}(\theta) &= \beta J \cos(\theta_{x_1} - \theta_{x_2}) \\ &= \frac{\beta J}{2} (\varphi_1(\theta_{x_1}) \varphi_{-1}(\theta_{x_2}) + \varphi_{-1}(\theta_{x_1}) \varphi_1(\theta_{x_2})) \\ \theta &= \{\theta_x, x\in Z^{\nu+1}\} \end{aligned} \tag{17}$$

(iii) *Lattice model of Young-Mills gauge field* (with compact gauge group G). The space of values $X = G$, v_0 is the normalized Haar measure in G. The gauge field $g = \{g_\tau\}$ is originally specified in the set of all oriented lines of lattice $Z^{\nu+1}$, in such a way that we have $g_\tau = (g_{-\tau})^{-1}$

(where $-\tau$ is the link resulting from a change of orientation of link τ). However, if we fix the values

$$g_\tau = e \text{ (unity of group G)} \tag{18}$$

for all links τ, parallel to the time axis (the so-called radial or Coulomb gauge), then the field is actually specified in the set $E \times Z^1$, where E is the set of all oriented links of lattice $Z^v = Y_0$. The supports of potential ϕ are any sets $A = \{\tau_1, \tau_2, \tau_3, \tau_4\}$ of links, forming a circuit of any two-dimensional face P of lattice Z^{v+1}, while the potential itself is then equal to

$$\phi_P = \beta \operatorname{Re} \chi_0 (g_{\tau_1} \cdot g_{\tau_2} \cdot g_{\tau_3} \cdot g_{\tau_4}) \tag{19}$$

where χ_0 is the character of a fixed irreducible representation $g \to T^0{}_g$ of group G (it is easily seen that ϕ_P is independent of the direction and origin of the circuit of face P over links τ_1, τ_2, τ_3, τ_4, and hence the notation ϕ_P is correct). The Gibbs field, based on potential (19), is called the Young-Mills lattice gauge field.

The basis $\{\varphi_\gamma\}$ in G is chosen as follows. Let $\{g \to T^\alpha_g\}$ be the set of all irreducible (unitary and finite-dimensional) representations of group G, acting respectively in spaces R^α. If we introduce into each R^α an orthonormalized basis $\{\eta^\alpha{}_m,\ m = 1, \ldots, \dim R^\alpha\}$, then the matrix elements of the representations

$$\varphi^\alpha{}_{m,m'} (g) = \frac{1}{(\dim R^\alpha)^{1/2}} (T^\alpha_g \, \eta^\alpha{}_m, \eta^\alpha{}_{m'}) \tag{20}$$

are well known to form an orthonormalized (with respect to the Haar measure) basis of functions in group G. The potential ϕ_P given by (19) is easily written as a sum of monomials (14) of the functions (20). We assume that there exists in group G an "elementary" representation $g \to \hat{T}_g$ (the same as its contragradient representation) such that every irreducible representation $g \to T^\alpha_g$ is contained as an irreducible component in some tensor power of the elementary representation. The least of these powers will in fact be called the rank of representation $g \to T^\alpha_g$, and accordingly, the rank of the function (20). It can easily be seen that relation (13) and estimate (13^a) hold here.

Note. We will mention an important feature of gauge fields. A gauge transformation in the space of field configurations (in radial gauge) will be defined as the transformation

$$(\tau_\gamma g)_\tau = \gamma(t_1)\, g_\tau\, \gamma^{-1}(t_2) \tag{21}$$

where t_1 and t_2 are the start and end of link τ, and $\gamma = \{\gamma(t), t \in Z^{\nu+1}\}$ is a finite function on lattice $Z^{\nu+1}$ with values in G and constant along the time direction: $\gamma(t + ne_{\nu+1}) = \gamma(t)$, $t \in Z^{\nu+1}$. It can easily be seen that a gauge transformation does not change potential (19), a radial gauge, or consequently, a Gibbs gauge Young-Mills field. The subspace $H^{g,inv} \subset H$, consisting of gauge-invariant functionals (i.e., invariant under transformations (21)) is invariant with respect to transfer matrix T, and its part $T^{g.inv}$ in $H^{g.inv}$ is usually called the gauge-invariant part of the transfer matrix. The most interesting problems arising in physics are concerned with spectral analysis of the operator $T^{g.inv}$.

3. Cluster Expansion of the Transfer Matrix

3.1. General Scheme of the Cluster Expansion of the Transfer Matrix

Let us mention here a basis in physical space H, in which the matrix elements of the transfer matrix can be written in a special way, which is important for what follows. The basis in question is obtained by a special orthogonalization of the (nonorthogonal) basis in

$$\varphi_\Gamma(\xi) = \prod_{x \in Y_0} \varphi_{\gamma(x)}(\xi_x) \tag{1}$$

where ξ is the configuration of the zero layer, $\{\varphi_\gamma, \gamma \in M\}$ is the basis of functions in X, mentioned in section 2.2, and $\Gamma = \{\gamma(x), x \in Y_0\}$ is a multiindex. In fact, we construct below the orthonormalized basis in H of multiplicative type

$$\psi_\Gamma(\xi) = \prod_{x \in Y_0} \psi^{(x)}_{\gamma(x)}(\xi) \tag{2}$$

where $\psi^{(x)}_\gamma$ are quasi-local functionals of the field configuration on layer Y_0 ($\psi^{(x)}_\gamma$ basically depend on the field values at points close to point x; this is expressed more precisely by relation (21)), which are such that, for all $s \in Z^\nu$,

$$\psi^{(x+s)}_\gamma(\xi) = \psi^{(x)}_\gamma(\tau_{-s}\,\xi). \tag{2^a}$$

It will be convenient for us to assume that the functions ψ_Γ depend on the entire field configurations in the lattice $Z^{\nu+1}$, and to introduce the identical system of functions

$$\hat{\psi}_{\Gamma} = U_{e_{v+1}} \psi_{\Gamma} = \prod_{x \in Y_1} \hat{\psi}^{(x)}_{\gamma(x)}; \; \hat{\psi}^{(x)}_{\gamma}(\xi) = \psi^{(x-e_{v+1})}_{\gamma}(\tau_{-e_{v+1}} \xi)$$

where $\Gamma = \{\gamma(x), x \in Y_1\}$ is a multiindex, defined on layer Y_1 and obtained by a shift of the multiindex $\Gamma = \{\gamma(x), x \in Y_0\}$ onto the zero layer Y_0. In this notation, the matrix elements $a_{\Gamma,\Gamma'} = (T\psi_{\Gamma}, \psi_{\Gamma'})$ of transfer matrix T in basis (2) can be written, in view of (6.2), in the form

$$\begin{aligned} a_{\Gamma,\Gamma'} &= < \hat{\psi}_{\Gamma} \, . \, \bar{\psi}_{\Gamma'} >_{\mu} \\ &= < \prod_{x \in Y_1} \hat{\psi}^{(x)}_{\gamma(x)} \, . \, \prod_{x' \in Y_0} \bar{\psi}^{(x')}_{\gamma'(x')} >_{\mu} \end{aligned} \tag{3}$$

On using the expression for the mean of a product of random variables in terms of their semi-invariants (see e.g. [35]), we finally obtain

$$a_{\Gamma,\Gamma'} = \Sigma \, \omega_{\Gamma_1,\Gamma_1'} \cdot \ldots \cdot \omega_{\Gamma_K,\Gamma_K'} \tag{4}$$

The summation here is over all divisions of the pair of sets

$$(T, T') = (\operatorname{supp} \Gamma, \operatorname{supp} \Gamma'), \; T \subset Y_1, \; T' \subset Y_0$$

into (unordered) sets of pairs $\{(T_1, T_1'), \ldots, (T_K, T_K')\}$ of nonempty subsets $T_i \subseteq T$, $T_i' \subseteq T'$, $i = 1, \ldots, K$, $K = 1, 2 \ldots$

Here,

$$\Gamma_i = \Gamma|_{T_i}, \quad \Gamma_i' = \Gamma'|_{T_i'}$$

are the contractions of Γ and Γ' onto these subsets. For any multiindices Γ and Γ' we have denoted by

$$\omega_{\Gamma,\Gamma'} = < \prod_{x \in Y_1}{}' \hat{\psi}^{(x)}_{\gamma(x)}, \; \prod_{x' \in Y_0}{}' \bar{\psi}^{(x')}_{\gamma'(x')} > \tag{5}$$

the semi-invariant of the system of functions (see [35])

$$\{\hat{\psi}^{(x)}_{\gamma(x)}, \bar{\psi}^{(x')}_{\gamma'(x')}, \; x \in \operatorname{supp} \Gamma, \; x' \in \operatorname{supp} \Gamma'\}$$

Note. Since the moments $< \psi_{\Gamma} > = < \prod_{x \in Y_0} \psi^{(x)}_{\gamma(x)} > = 0$ for any multiindex Γ, all the semi-invariants $< \prod'_{x \in Y_0} \psi^{(x)}_{\gamma(x)} >$ are also equal to zero. This is why there appear in expansion (4) only divisions into pairs with nonempty sets T_i and T'_i. It follows from (2^a) that, for all $s \in Z^v$,

$$\omega_{\Gamma+s,\Gamma'+s} = \omega_{\Gamma,\Gamma'}, \tag{5^a}$$

where $\Gamma + s$ and $\Gamma' + s$ are the multiindices obtained by shifts of Γ and Γ' by s.

Below, when basis (2) is constructed explicitly, we shall see that, for many potentials of type (14.2), given sufficiently small values of β, the semi-invariants $\omega_{\Gamma,\Gamma'}$ satisfy the so-called cluster estimate

$$|\omega_{\Gamma,\Gamma'}| \leq L(\lambda_1(\beta))^{d_{T\cup T'}} \prod_{x\in Y_1} (\lambda_2(\beta))^{\ell(\gamma(x))} \prod_{x'\in Y_0} (\lambda_2(\beta))^{\ell(\gamma'(x'))} \quad (6)$$

where L is a constant (independent of β), $\lambda_1(\beta)$ and $\lambda_2(\beta)$ are functions of β such that $\lambda_{1,2}(\beta) \to 0$ as $\beta \to 0$,

$$T = \text{supp}\,\Gamma,\ T' = \text{supp}\,\Gamma',$$

d_B for $B \subset Z^{v+1}$ is the length of the least tree constructed on all points B, and $\ell(\gamma)$ is the rank of index γ (see above). Expansion (4) together with estimate (6) of semi-invariant $\omega_{\Gamma,\Gamma'}$ is known as the *cluster expansion* of the transfer matrix. This expansion leads us to the definition of a general class of operators which are of interest in themselves.

We denote by M the set of all multiindices $\Gamma = \{\gamma(x), x\in Y_0\}$; let $\ell_2(M)$ be the Hilbert space of functions $f = \{f(\Gamma), \Gamma\in M\}$ with finite norm

$$\|f\| = (\sum_{\Gamma\in M} |f(\Gamma)|^2)^{1/2} < \infty$$

We call operator A, acting in $\ell_2(M)$ according to the relation

$$(Af)(\Gamma) = \sum_{\Gamma'\in M} a_{\Gamma,\Gamma'} f(\Gamma') \quad (7)$$

a *cluster* operator, if its matrix elements $a_{\Gamma,\Gamma'}$ have the form

$$a_{\Gamma,\Gamma'} = \Sigma\, \omega_K [(\Gamma_1, \Gamma_1'), \ldots, (\Gamma_K, \Gamma_K')] \quad (8)$$

where the summation is over all the divisions as in (4), while ω_k are functions (cluster functions) defined in all unordered sets of k pairs of multiindices

$$\{(\Gamma_1, \Gamma_1'), \ldots, (\Gamma_K, \Gamma_K')\},\ K = 1, 2, \ldots \quad (9)$$

(multiindices Γ_i are defined as above on layer Y_1) and satisfy the conditions:

1. given any set of pairs (9) and set of vectors of lattice Z^v

$$(s_1, \ldots, s_K)$$

$$\omega_K [(\Gamma_1 + s_1, \Gamma_1' + s_1), \ldots, (\Gamma_K + s_K, \Gamma_K' + s_K)] = \omega_K [(\Gamma_1, \Gamma_1'), \ldots, (\Gamma_K, \Gamma_K')] \quad (10)$$

where $\Gamma + s$ is the "shifted" multiindex;
2. we have

$$|\omega_K [(\Gamma_1, \Gamma'_1), \ldots, (\Gamma_K, \Gamma'_K)]| <$$
$$< \prod_{i=1}^{K} (\lambda_1^{d_{T_i \cup T'_i}} \cdot \prod_{x \in Y_1} \lambda_2^{\ell(\gamma_i(x))} \cdot \prod_{x' \in Y_0} \lambda_2^{\ell(\gamma'_i(x'))}) \tag{11}$$

where T_i, T'_i are supports of multiindices Γ_i and Γ'_i respectively, λ_1 and λ_2 are numbers, $0 < \lambda_1 < 1$, $0 < \lambda_2 \leq 1$, called the *clustering parameters*.

Provided that relations (5^a) and (6) are satisfied, the transfer matrix T (regarded as an operator in $\ell_2(M)$) is a cluster operator of a special type, in which the k-th cluster function is obtained as a k-tuple product of the first cluster function. The cluster operators with this property are usually known as *multiplicative*.

Note. When the marking set M consists of two elements (θ, 1), any multiindex Γ can be identified with a set $T = \text{supp}\,\Gamma$, and the cluster operator can be regarded as acting in the space $\ell_2(C_{Z^\nu})$ of functions defined in finite subsets Z^ν. On the right-hand side of estimate (11) there then remain only factors of the type $\lambda_1^{d_{T_i \cup T'_i}}$ (i.e., it can be assumed that $\lambda_2 = 1$).

3.2. *Construction of a Multiplicative Basis and Derivation of Cluster Estimates*

Here we shall explicitly construct multiplicative basis (2) for the case of potential (14.2). For simplicity, we will assume that the interaction radius of the potential $r_0 = 1$ (interaction of "nearest neighbors"), and that the rank of the potential is unity (i.e. the monomials φ_Γ in expansion (14.2) of the potential ϕ contain factors φ_γ of rank $\ell(\gamma) = 1$).

We introduce the lexicographic ordering of points of the layer Y_0, and for every point $x_0 \in Y_0$ we denote by $< \,.\, | \xi_{x<x_0} = \xi >_\mu$ the mean over the conditional distribution $\mu(\,.\,|\xi_{x<x_0} = \xi)$, generated by Gibbs measure μ, provided that the field values $\xi_{x<x_0} = \xi$ in the "past" with respect to x_0 are fixed. For $\gamma \neq \theta$ we put

$$\tilde{\varphi}_\gamma^{(x_0)}(\xi) = \varphi_\gamma(\xi_{x_0}) - <\varphi_\gamma(\xi_{x_0}) | \xi_{x<x_0} = \xi>_\mu \tag{12}$$

and let

$$G_{\gamma,\gamma'}^{(x_0)}(\xi) = <\tilde{\varphi}_\gamma^{(x_0)} \,.\, \bar{\tilde{\varphi}}_{\gamma'} | \xi_{x<x_0} = \xi>_\mu \tag{13}$$

be the elements of the conditional Gram matrix $\{G^{(x_0)}_{\gamma,\gamma'}\}$ for the functions $\{\varphi^{(x_0)}_\gamma\}$.

Using the cluster expansion of the conditional Gibbs field, along with (13.2) and estimate (13ª.2), it can be shown that, for the "nearest neighbors" potentials of type (14.2) and rank 1, we have the following expansion (for more details see [20.II]):

$$<\varphi_\gamma(\xi_{x_0}) \mid \xi_{x<x_0} = \xi_x> = \sum_{\Gamma:\ \mathrm{supp}\,\Gamma<x_0} T_\Gamma^{(x_0,\gamma)}\ \varphi_\Gamma(\xi) \tag{14}$$

in which the coefficients $T_\Gamma^{(x_0,\gamma)}$ admit of the estimate

$$|T_\Gamma^{(x_0,\gamma)}| < (\lambda_1(\beta))^{d_{\{x_0\}\cup \mathrm{supp}\,\Gamma}}\ (\lambda_2(\beta))^{\ell(\gamma)-1} \prod_{x\in \mathrm{supp}\,\Gamma} \lambda_2(\beta)^{\ell(\gamma(x))-1} \tag{15}$$

where $\lambda_i(\beta) = C_i\beta$, $i=1, 2$, C_1, C_2 are constants, independent of β.

A similar expansion also holds for $G^{(x_0)}_{\gamma,\gamma'}(\xi)$:

$$G^{(x_0)}_{\gamma,\gamma'}(\xi) = \delta_{\gamma,\gamma'} + \sum_{\Gamma:\ \mathrm{supp}\,\Gamma<x_0} S_\Gamma^{(x_0,\gamma,\gamma')}\ \varphi_\Gamma(\xi) \tag{16}$$

Here,

$$|S_\Gamma^{(x_0,\gamma,\gamma')}| < (\lambda_1(\beta))^{d_{\{x_0\}\cup \mathrm{supp}\,\Gamma}}\ (\lambda_2(\beta))^{|\ell(\gamma)-\ell(\gamma')|} \prod_{x\in \mathrm{supp}\,\Gamma} \lambda_2(\beta)^{\ell(\gamma(x))-1} \tag{17}$$

It follows from expansion (16) and estimate (17) that the elements $B^{(x_0)}_{\gamma,\gamma'}(\xi)$ of the matrix $B^{(x_0)} = [G^{(x_0)}]^{-1/2}$ have the form

$$B^{(x_0)}_{\gamma,\gamma'}(\xi) = \delta_{\gamma,\gamma'} + \sum_{\Gamma:\ \mathrm{supp}\,\Gamma<x_0} M_\Gamma^{(x_0,\gamma,\gamma')}\ \varphi_\Gamma(\xi) \tag{18}$$

where the coefficients $M_\Gamma^{(x_0,\gamma,\gamma')}$ admit of an estimate of type (17). The matrix $B^{(x_0)} = \{B^{(x_0)}_{\gamma,\gamma'}\}$ conditionally norms the system of functions $\{\tilde\varphi_\Gamma^{(x_0)}\}$, i.e., the functions

$$\psi_\gamma^{(x_0)} = \sum_{\gamma':\ \gamma'\neq\theta} B^{(x_0)}_{\gamma,\gamma'}\ \tilde\varphi_{\gamma'}^{(x_0)} \tag{19}$$

as functions of the variable ξ_{x_0}, form an orthonormalized basis in X with respect to the conditional distribution of values of the field ξ_{x_0} at the point x_0 under the condition that the values of field $\xi_{x<x_0} = \xi$ are fixed. Hence it is easily seen that the system of functions in

$$\psi_\Gamma(\xi) = \prod_{x\in Y_0} \psi^{(x)}_{\gamma(x)} \tag{20}$$

is orthonormalized. This system proves to be complete in H (see [29]).

Notice that, by the above expansions and estimates, the function $\psi_\gamma^{(x)}$ can be written as

$$\psi_\gamma^{(x)}(\xi) = \varphi_\gamma(\xi_x) + \sum_{\Gamma:\, \mathrm{supp}\,\Gamma \leqslant x} R_\Gamma^{(x,\gamma)} \varphi_\Gamma \tag{21}$$

where the coefficients $R_\Gamma^{(x,\gamma)}$ admit of estimates of type (15). Using expansion (21) together with the estimates for the coefficients $R_\Gamma^{(x,\gamma)}$, and the general methods developed in [35] for estimating the semi-invariants for quasi-local functionals of a Gibbs field, we arrive at estimate (6).

Notes. 1. If the potential ϕ has an arbitrary interaction radius r_0 and arbitrary rank ℓ_0, we have to put the functions $\lambda_1(\beta)$ and $\lambda_2(\beta)$ in the above estimates equal to

$$\lambda_1(\beta) = C_1\, \beta^{1/z0},\ \lambda_2(\beta) = C_2\, \beta^{1/\ell_0}$$

where C_1, C_2 are constants.

2. The scheme for constructing the multiplicative basis can be traced back to Minlos and Sinai [1], and was subsequently used in [2–4], [6], [10–16], [18–20], [26], [29], and [33]. The idea of asymptotic multiplicativeness of the transfer matrix likewise appeared in [1]. In Malyshev [2] and Abdulla-Zadeh, Minlos, and Pogosian [4] the idea was clearly stated as the cluster expansion (4), while the general concept of a cluster operator was introduced for the first time in [4].

3. In the case of lattice gauge-invariant Young-Mills fields with Abelian gauge group, Khrapov constructed in [19] a multiplicative basis, similar to basis (2), in the actual physical space of gauge-invariant functionals, and he obtained the cluster expansion (4) of the gauge-invariant part of the transfer matrix.

3.3. Some Generalizations

In the previous sections we have described the cluster expansion of a transfer matrix of a Gibbs random Markov field in the case of small β (i.e., Gibbs disturbance of an independent field). We shall here indicate two further situations in which similar results can be obtained.

Gibbs field at low temperatures (large β)

We take the spin lattice model with interaction of nearest neighbours on lattice $Z^{\nu+1}$ (the Ising model with zero magnetic field). For this model in the case of large $\beta \gg 1$ there are two limiting Gibbs distributions (in the sense of the DLR definition, see Para. 1.2). For one of them, configurations $\sigma = \{\sigma_x, x \in Z^{\nu+1}\}$ with a preponderance of values $\sigma_x = +1$ (the

(+)-phase) are typical, and for the other, configurations with a preponderance of values $\sigma_x = -1$ (the (−)-phase). For every configuration σ we denote by $\Gamma(\sigma)$ its boundary, i.e., the set of v-dimensional faces of the shifted lattice $\tilde{Z}^{v+1}$ which separate the adjacent points x, $x' \in Z^{v+1}$ with different values of the field; the connected components $\Gamma(\sigma)$ will be called contours of configuration σ. For large β, each configuration (± is the phase) generates an "ensemble of contours," i.e., a probability distribution $P_\pm$ on a set $\sigma\iota^{v+1}$ of configurations $\alpha = \{\Gamma_i\}$ of finite contours on $\tilde{Z}^{v+1}$, in which each contour Γ_i is embraced by only a finite number of other contours of configuration α (we have $P_+ = P_-$ in the case of the Ising model, but for more general spin models these distributions may be distinct, see Sinai [46]). Thus, when describing a field with large β, we can consider Hilbert spaces $L_2(\sigma\iota^{v+1}, P_\pm)$ of functionals of configurations of contours on $\tilde{Z}^{v+1}$; the physical space $H_\pm$ is then the same as the space of functionals on $\sigma\iota^{v+1}$, which depend, not on the entire configuration α, but only on its section by the zero layer Y_0. In the case $v>1$ every such section is again a configuration of (v-dimensional) contours on $\tilde{Z}^v$, while with $v=1$ the section can be described as a "denumerable" set $C \subset Z^v = Y_0$ (over which the contours of configuration α intersect Y_0) together with its two divisions into a pair of points, namely, the "past" division π^{pst} (in which those points of C which are connected by a piece of some contour $\Gamma \in \alpha$, lying wholly in the past are combined in one pair), and the similar "future" division π^{fut}. (The future and past divisions are naturally matched.) Obviously, given a fixed section of configuration α by the zero layer Y_0, the conditional ensembles of contours in the "past" and "future" are independent (the Markov property). We shall henceforth confine ourselves to the case $v=1$ and choose in space H the (not orthonormalized) basis

$$\chi_{\{\tau_i^{pst}\},\{\tau_j^{fut}\}}(C, \pi^{pst}; \pi^{fut}) = \prod_i \chi_{\tau_i^{pst}} \cdot \prod_j \chi_{\tau_j^{fut}} \qquad (22)$$

where $\{\tau_i^{pst}\}$ and $\{\tau_j^{fut}\}$ are the two sets of pairs of points (pairs in each set do not intersect), while

$$\chi_{\tau^{pst}}(C, \pi^{pst}, \pi^{fut}) = \begin{cases} 1 \text{ if the pair } \tau^{pst} \subset C \text{ and is a pair of a past division,} \\ 0 \text{ otherwise.} \end{cases}$$

We similarly define $\chi_{\tau^{fut}}$.

The system $\{\chi_{\tau^{pst}}\}$ and $\{\chi_{\tau^{fut}}\}$ can then be subjected to orthogonalization, in the same way as we described in the previous section; then we can construct a multiplicative orthonormalized basis in H, in which the matrix elements of transfer matrix T admit of cluster expansion, similar to expansion (4). Unfortunately, in the case of the Ising model this program encounters certain combinatorial difficulties in obtaining the required entropy estimates; this is bound up with the large variety of possible shapes of contours. In the case of simplified spin models, which allow only contours of simple shape (e.g., only rectangular contours, or rectangular contours with a small number of "teeth"), Malyshev, Minlos, and Khrapov constructed in [13] a multiplicative basis, as described above, and obtained the cluster expansion of the matrix elements of the transfer matrix in this basis. We shall return to this case when discussing the spectrum of the transfer matrix.

Fermi quasi-states

It turns out to be possible to define and study the transfer matrix, not only for classical random fields, but also in the case of so-called "super-fields" (i.e., states on quasi-loçal super-algebras, see e.g., [35]).

Let $\mathfrak{A} = \mathfrak{A}_{\nu+1}$ be a Grassmann algebra with the system of generators $\{\psi_x^\alpha, \bar{\psi}_x^\alpha, x\in Z^{\nu+1}, \alpha = 1, \ldots, s\}$. Let $\phi = \{\phi_A, A\subset Z^{\nu+1}\}$ be the potential, i.e., a family of even polynomials of generators (there are only generators ψ_x^α and $\bar{\psi}_x^\alpha$ with $x\in A$ in ϕ_A). Then we define a quasi-state (a linear normed functional) in $\mathfrak{A}$ by

$$<F_B> = \lim_{\Lambda \nearrow Z^{\nu+1}} <F_B \exp\{\sum_{A\subseteq\Lambda} \phi_A\}>_{0,\Lambda} \qquad (23)$$

where F_B, $B\subset Z^{\nu+1}$ is a local element of $\mathfrak{A}$ (polynomial of ψ_x^α and $\bar{\psi}_x^\alpha$, where $x\in B$), called the Gibbs (Fermion) quasi-state in $\mathfrak{A}$, generated by potential ϕ; here, $< >_{0,\Lambda}$ is the so-called 'Berezin integral", i.e., the simplest quasi-state in algebra $\mathfrak{A}_\Lambda \subset \mathfrak{A}$ (see [35, 47]). If, e.g., the interaction $\sum_{A\subseteq\Lambda} \phi_A$ in (23) is a "small disturbance" of the quadratic interaction $u^0_\Lambda = \sum_{\alpha, x\in\Lambda} \psi_x^\alpha \bar{\psi}_x^\alpha$ (see Kashanov and Malyshev [14]), the limit (23) exists and satisfies the so-called Ostervalder-Schrader positiveness condition, see [54],

$$<F . (\theta F)> \geqslant 0 \qquad (24)$$

for any element $F\in\mathfrak{A}^+$; here, $\mathfrak{A}^+$ is the subalgebra generated by generators of the "future," while θ: $\mathfrak{A}^+ \to \mathfrak{A}^-$ is the natural anti-linear homomorphism of algebra $\mathfrak{A}^+$ into algebra $\mathfrak{A}^-$, generated by the "past." On augmenting $\sigma \iota^+$ up to the semi-scalar product

$$(F_1, F_2) = \, <F_1 \cdot (\theta F_2)>$$

and factoring this augmentation with respect to the kernel $N = \{F \in \sigma \iota^+:$ $(F, F) = 0\}$, we construct the physical Hilbert space H and the transfer matrix in it, induced by the natural homomorphism

$$U_{e_{v+1}}: \sigma \iota^+ \to \sigma \iota^+: \psi_x^\alpha \to \psi_{x+e_{v+1}}^\alpha, \, \overline{\psi}_x^\alpha \to \overline{\psi}_{x+e_{v+1}}^\alpha.$$

In the above-mentioned paper [14], Kashapov and Malyshev used a modification of the above scheme to construct a multiplicative basis in space H and obtained the cluster expansion of the transfer matrix.

4. Invariant Cluster K-Particle Subspaces of the Transfer Matrix

4.1. Definitions

The first step in studying the leading branches of the spectrum of the transfer matrix T (or what amounts to the same thing, the lower branches of the spectrum of the corresponding Hamiltonian, see the Introduction) is to isolate certain very simple subspaces in H, invariant under T and the group of translations $\{U_s, s \in Z^v\}$ of the zero layer Y_0. The subspace $H_1 \subset H$, invariant under T and cyclical with respect to group U_s (i.e., generated by the vectors $U_s h_0$, where h_0 is a fixed vector), is usually called the one-particle invariant subspace. When the vector h_0 can be chosen in such a way that the vectors $h_s = U_s h_0$ form an orthonormalized basis in H_1, and the matrix elements

$$(Th_s, h_{s'}) = a_{s-s'} \tag{1}$$

decrease exponentially with the distance between s and s'

$$|a_{s-s'}| < C \lambda^{|s-s'|} \tag{1a}$$

where $0 < \lambda < 1$ and $C > 0$ are constants, then we agree to call the subspace H_1 the *cluster one-particle* subspace. In the more general case we consider the *marked cluster one-particle* invariant subspace $H_1 \subset H$: in it there exists the orthonormalized basis $\{h_x^\gamma, x \in Y_0, \gamma \in M\}$, marked by the points x and by the elements of a denumerable or finite set M (in a similar way to that described in section 2.2), such that

$$U_s h_x^\gamma = h_{x+s}^\gamma, \, s \in Z^v \tag{2}$$

and the matrix elements of the transfer matrix

$$(Th_x^\gamma, h_{x'}^{\gamma'}) = a_{x-x'}^{\gamma,\gamma'}$$

have the bound

$$|a_{x-x'}^{\gamma,\gamma'}| < C\lambda_1^{|x-x'|}\lambda_2^{\ell(\gamma)+\ell(\gamma')} \tag{3}$$

for $|x-x'| > 1, 0 < \lambda_i < 1, i=1,2$, C is a constant, and $\ell(\gamma)$ is the rank of index γ.

More complex invariant subspaces are the cluster two-particle subspaces $H_2 \subset H$; in them there exists a basis $\{h_{\{x_1,x_2\}}, \{x_1, x_2\} \subset Y_0\}$, marked by unordered pairs $\{x_1, x_2\}$ of points of the lattice Z^ν (or sometimes only by pairs of distinct points, i.e., two-point sets of Z^ν). It is then required that

$$U_s h_{\{x_1,x_2\}} = h_{\{x_1+s, x_2+s\}} \tag{4}$$

and that the matrix elements of transfer matrix $a_{\{x_1,x_2\},\{x_1',x_2'\}} = (T h_{\{x_1,x_2\}}, h_{\{x_1',x_2'\}}$ in this basis admit of the cluster expansion

$$\begin{aligned} a_{\{x_1,x_2\},\{x_1',x_2'\}} &= \omega_1(x_1-x_1', x_2-x_2') + \\ &+ \omega_1(x_1-x_2', x_2-x_1') + \omega_2(x_1, x_2; x_1', x_2') \end{aligned} \tag{5}$$

the cluster functions ω_1 and ω_2 being such that

$$|\omega_1(y_1, y_2)| < C\lambda^{|y_1|+|y_2|} \tag{6}$$

and

$$\begin{aligned} &|\omega_2(x_1, x_2; x_1', x_2')| < C\lambda^{d_{\{x_1+e_{\nu+1}, x_2+e_{\nu+1}, x_1', x_2'\}}} \\ &\omega_2(x_1+s, x_2+s; x_1'+s, x_2'+s) \\ &= \omega(x_1, x_2, x_1', x_2')\ s \in Z^\nu \end{aligned} \tag{7}$$

where C and λ are constants, $0<\lambda<1$.

In the more general case, marked two-particle cluster subspaces appear, in which there is an orthonormalized basis $\{h_\Gamma\}$, marked by multiindices Γ with values in M and such that $|\mathrm{supp}\,\Gamma| = 2$. The group $\{U_s\}$ acts in this basis in the same way as (4), while the matrix elements of the transfer matrix admit of expansion similar to expansion (5), together with estimates of the cluster functions similar to estimates (II.3).

In a similar way we can define k-particle invariant cluster subspaces of T: in them there is a basis $h_{\{x_1,\ldots,x_k\}}$, marked by unordered sets of points of the lattice Z^ν (either by all such sets, or by sets of pairwise distinct points, i.e., by k-point subsets of Z^ν). Here,

$$U_s h_{\{x_1,\ldots,x_k\}} = h_{\{x_1+s,\ldots,x_k+s\}} \tag{8}$$

while the matrix elements of the transfer matrix in this basis admit of a cluster expansion similar to (8.3), with estimates (II.3), (in which $\lambda_2 = 1$). Marked k-particle invariant subspaces of the transfer matrix may be defined in a similar way.

4.2. Construction of K-Particle Invariant Subspaces

1. Spin fields

The fullest results on the existence of k-particle cluster invariant subspaces have been obtained in the case of the transfer matrix of a spin field (see Minlos and Sinai [1], and Malyshev and Minlos [6]). In this case the marking set M consists of the two elements $(\theta, 1)$ (see section 3.2, Example 1), and hence unmarked k-particle subspaces arise here.

For simplicity consider the case of the (pair) potential (16.2):

$$\begin{gathered} \phi_{\{x_1,x_2\}} = \beta\, J_{x_1-x_2}\, \sigma_{x_1}\, \sigma_{x_2}, \quad x_1, x_2 \in Z^{\nu+1} \\ J_y = 0 \text{ for } |y^{\nu+1}| > 1 \end{gathered} \tag{9}$$

and assume that the function

$$j_s = J_{s+e_{\nu+1}}\ s \in Z^\nu$$

is either positive-definite or negative-definite, i.e.,

$$\Big|\sum_{x,\,s\in Z^\nu} j_s\, Z_x\, \overline{Z}_{x+s}\Big| > C_0 \sum_{x\in Z^\nu} |Z_x|^2 \tag{10}$$

for any sequence $\{Z_x, x \in Z^\nu\} \in \ell_2(Z^\nu)$, $C_0 > 0$ is a constant.

It turns out that, in the case of transfer matrix T of the Gibbs field, which is based on (pair) potential (9) and satisfies condition (10), for small values of β, there exist $N = N(\beta)$ invariant subspaces H_k, $k = 1, \ldots, N$, which are one-, two-, . . ., N-particle cluster subspaces such that the spectrum $\sigma(T|_{H_k})$ of the corresponding part of the transfer matrix lies in the range

$$\sigma(T|_{H_k}) \subset [C_1\beta^k, C_2\beta^k] \tag{11}$$

where $C_1 < C_2$ are constants. In the orthogonal complement to $\bigoplus_{k=0}^{N} H_k$ (H_0 is the space of constants), the spectrum of T lies in the range $[-C_3\beta^{N+1}, C_3\beta^{N+1}]$, where C_3 is constant.

Notes. (i) This result also holds for any unmarked cluster operator in $\ell_2(C_{Z^\nu})$ with sufficiently small clustering parameter λ_1, for which the values of the cluster functions for the two-point pairs satisfy the following condition: the functions

$$j_k(s_1, \ldots, s_k) = \omega_k[(\{0\}, \{s_1\}), \ldots, (\{0\}, \{s_k\})] \qquad (12)$$

are sign-definite for any k.

(ii) Notice that, for sufficiently small β, it follows from (11) that the **spectra** $\sigma(T|_{H_k})$ in different subspaces H_k, $k = 1, \ldots, N$, do not intersect. A posteriori, this proves to be the main reason why some first invariant subspaces of the transfer matrix T can be isolated. When β is fixed and k is large, the spectra in the invariant k-particle subspaces (if there are any such) start to overlap, producing the main difficulty for isolating these subspaces.

Let us briefly describe a device for constructing the invariant k-particle subspaces of the transfer matrix of a spin field (or more generally, of a cluster operator). We shall confine ourselves to the case $k = 1$. We divide the space $\ell_2(C_{Z^\nu}^{\geq 1})$ into the direct sum

$$\ell_2(C_{Z^\nu}^{\geq 1}) = \ell_2(Z^\nu) \oplus \ell_2(C_{Z^\nu}^{\geq 2}),$$

where $C_{Z^\nu}^{\geq s}$ is the set of finite subsets $T \subset Z^\nu$, $|T| \geq s$. The cluster operator A can here be written as the operator matrix

$$A = \begin{pmatrix} A_{11} & A_{12} \\ A_{21} & A_{22} \end{pmatrix}$$

where $A_{11}: \ell_2(Z^\nu) \to \ell_2(Z^\nu)$, $A_{12}: \ell_2(C_{Z^\nu}^{\geq 2}) \to \ell_2(Z^\nu)$, etc. We shall next seek the one-particle subspace H_1 as a disturbance of the space $\ell_2(Z^\nu)$ of the type

$$H_1 = \{f + Sf, f \in \ell_2(Z^\nu)\}, \qquad (13)$$

where S is the operator: $S: \ell_2(Z^\nu) \to \ell_2(C_{Z^\nu}^{\geq 2})$.

The requirement that H_1 be invariant with respect to A is equivalent to the relation

$$(A_{21} + A_{2s}S) = S(A_{11} + A_{12}S),$$

which can be rewritten as

$$S = A_{21}A_{11}^{-1} + A_{22}S\,A_{11}^{-1} - SA_{12}\,SA_{11}^{-1} \qquad (14)$$

and considered as an equation in S. By means of cluster estimates (II.3) for A and estimate (12), it can be shown that the right side of (14) specifies a contraction mapping in the space of cluster operators, acting from $\ell_2(Z^\nu)$ into $\ell_2(C_{Z^\nu}^{\geq 2})$ (suitably normed, for more details see [6]). Thus, given a small clustering parameter (small β), Eq. (14) is solvable, and its solution S (which automatically has a small norm) specifies an invariant subspace H_1 of type (13). Let

$$\{U_x = e_x + Se_x \in H_1, x \in Z^\nu\}.$$

The system of vectors in H_1 is $e_x(y) = \delta_{x,y} \in \ell_2(Z^\nu)$.
Their Gramm matrix is

$$G_{x,x'} = (U_x, U_{x'}) = \delta_{x,x'} + (S^*S)_{x,x'}.$$

Since S is a cluster operator and has a small norm, the operator $G^{-1/2} = (E + S^*S)^{-1/2}$ in $\ell_2(Z^\nu)$ exists and can be written as

$$G^{-1/2} = E + T,$$

where the operator T is specified by the matrix $T_{x,x'} = t_{x-x'}$, while

$$|t_{x-x'}| < K\,(C\lambda)^{|x-x'|} \tag{15}$$

$K>0$, $C>0$, λ is the clustering parameter of operator A. Hence it follows that the vectors

$$h_x = U_x + \sum_{x' \in Z^\nu} t_{x-x'}\, U_{x'}$$

form an orthonormalized basis in H, where $U_s h_x = h_{x+s}$, and for the matrix elements $(Ah_x, h_{x'})$ of the operator $A|_{H_1}$, we have an estimate of type (15).

Two-, three-, . . ., k-particle cluster invariant subspaces of A can be similarly constructed. In these constructions and estimates, essential use is made of the following important lemma: the product of any number of cluster operators

$$A_1 \ldots A_m \tag{16}$$

(with the same clustering parameter λ) is again a cluster operator, for which the clustering parameter does not exceed $C\lambda$, where C is a constant, independent of the number of factors in (16).

2. Other fields

While the above device can also be used to construct k-particle subspaces of the transfer matrix in the case of Gibbs fields with more complicated spin space X, there can then arise several one-particle invariant subspaces in which the spectrum has the same order in β, and several two-particle subspaces with close spectra, etc. This fact makes it much more difficult to isolate and study these subspaces. Let us list the results so far known about the existence of k-particle subspaces for transfer matrices of Gibbs fields.

A. Fields with a compact space of spins. For such fields (see Malyshev [2], Zhelondek [15], Khrapov [20], Chernykh [26]), generated by a (pair) potential of the type (cf. (14.2))

$$\Phi_{\{x_1, x_2\}}(\xi) = \beta \Sigma \beta_\gamma \varphi_\gamma(\xi_{x_1}) \varphi_\gamma(\xi_{x_2}),$$

where $\{\varphi_\gamma, \gamma \in M\}$ is the basis described in Para. 2.2, β_γ are constants, and the summation is over a (finite) set $\epsilon \subseteq M$ of indices of rank $\ell(\gamma) = 1$, it can be shown that (for small β) there exists a unique marked cluster one-particle subspace H_1, in which there exists the basis $\{h_x^\gamma, x \in Z^\nu, \gamma \in \epsilon\}$, having properties (2) and (3). The spectrum of the transfer matrix T in H_1 has order β, while its spectrum in the orthogonal complement to H_1 has higher order in β. In certain cases, H_1 breaks up into the direct sum of unmarked one-particle invariant cluster subspaces with pairwise disjoint spectra (see Section 5).

B. The gauge-invariant part of the transfer matrix of the Young-Mills lattice field. Let G be a compact gauge group and let the Young-Mills interaction (19.2) be specified by the character χ_0 of an "elementary representation" $g \to T_g^0$ of group G (see Para. 3.2; for instance, T_g^0 is a vector representation in the case of groups O_n or SU(n)). In this case the gauge-invariant part of the transfer matrix $T^{\text{gau.inv}}$ has one one-particle cluster invariant subspace H_1 with spectrum $\sim \beta^4$ (basic glueballs). In the case when the contragradient representation $g \to (T_g^0)^*$ is the same as T_g^0, the space H_1 is unmarked (neutral glueball), while in the case when $(T_g^{0*} \neq T_g^0$, it is marked by two values $\gamma = \pm 1$ (glueball charge) (on this topic, see Abdulla-Zadeh [11, 12], Minlos and Khrapov [18], Khrapov [19], and Schor [40, II]).

Sherstneva [33] has constructed for group G = U(1) (circle) the next (in order of magnitude of spectrum) "excited" one-particle subspaces with spectrum $\sim \beta^6$ and $\sim \beta^8$ (excited glueballs), and also two-particle invariant subspaces (with spectrum $\sim \beta^8$).

C. Fermion quasi-state (see Section 3.3). For the transfer matrix of the

Fermion quasi-state described in Section 3.3 Kashapov and Malyshev constructed in [14] several leading cluster k-particle invariant subspaces, in a similar way to that described above in the case of a spin lattice field.

D. Spin fields at low temperature (ensemble of contours). In the case of the simplified two-dimensional contour models referred to in Section 3.3, Malyshev, Minlos and Khrapov showed in [13] that the leading (in order of magnitude of the spectrum) invariant subspace of the transfer matrix (the contour ensemble) is not a one-particle subspace, as is usually the case for small β, but is a two-particle subspace. In other words, in this case there are no single localized elementary excitations; instead, they appear in pairs, in which the excitations can move almost independently of one another (or possibly form "bound states.")

5. Cluster Operators in the "P-Representation"

5.1. Transition to a Fourier Transformation

For a more detailed study of the spectrum of the transfer matrix and its invariant n-particle subspaces, it is convenient to pass to the "Fourier transformation" of a function $f(x_1, \ldots, x_n)$ of the "lattice" variables $x_i \in Z^\nu$, $i=1, \ldots, n$:

$$f(x_1, \ldots, x_n) \to \tilde{f}(p_1, \ldots, p_n) = \sum_{\{x_1, \ldots, x_n\}} f(x_1, \ldots, x_n) \prod_{k=1}^{n} e^{i(p_k, x_k)}$$

where the functions $\tilde{f}(p_1, \ldots, p_n)$ are dependent on the variables $p_i \in T^\nu$ on the ν-dimensional torus, and then to consider the action of the cluster operators directly on the functions $\tilde{f}(p_1, \ldots, p_n)$. This leads to a very nice class of operators in the space of these functions, which we shall likewise refer to as cluster operators (or cluster operators in the "p-representation"). Let us describe this class.

Let $L_n = L_2((T^\nu)^n)$ be a Hilbert space of functions $f(p_1, \ldots, p_n)$ of n variables $p_i \in T^\nu$, varying on the ν-dimensional torus T^ν. The operator A, acting in L_n according to the relation

$$(Af)(p_1, \ldots, p_n)$$

$$= \int_{(T^v)^n} K(p_1, \ldots, p_n; q_1, \ldots, q_n) f(q_1, \ldots, q_n)\, dq_1 \ldots dq_n \qquad (1)$$

will be called a *cluster* operator if its (generalized) kernel K has the form

$$K(p_1, \ldots, p_n; q_1, \ldots, q_n) = \sum_{\gamma} a_\gamma (p_1, \ldots, p_n, q_1, \ldots, q_n)\delta_\gamma \quad (2)$$

where the summation is over all divisions γ of the pair of sets $N_n = \{1, \ldots, n\}$, $N_n' = \{1, \ldots, n\}$ into pairs of nonempty subsets:

$$\gamma = \{(\beta_1, \beta_1'), \ldots, (\beta_k, \beta_k')\}, \beta_i \subseteq N_n, \beta_i' \subseteq N_n' \qquad (3)$$

while

$$\delta_\gamma = \prod_{i=1}^{k} \delta(p_{\beta_i} - Q_{\beta_i'}) \qquad (4)$$

Here, given any subsets $\beta \subseteq N_n$ and $\beta' \subseteq N_n'$, we denote by P_β and $Q_{\beta'}$:

$$p_\beta = \sum_{i\in\beta} p_i, \; Q_{\beta'} = \sum_{i\in\beta'} q_i \qquad (5)$$

(addition is understood as addition in group T^v); $\delta(.)$ is a δ-function on the torus T^v (the measure dq on T^v is the normalized Haar measure). Further, a_γ is an analytic function on the manifold Γ_γ in $(T^v)^n \times (T^v)^n$:

$$\Gamma_\gamma = \{(p_1, \ldots, p_n; q_1, \ldots, q_n): P_{\beta_i} = Q_{\beta_i'}, i = 1, \ldots, k\} \qquad (6)^*$$

(γ is a division of type (3)), which will be called a cluster function of the operator A.

It is easily seen that the operator A given by (1) is bounded. In the case when

$$a_\gamma (p_1, \ldots, p_n; q_1, \ldots, q_n) = a_{\gamma^*} (q_1, \ldots, q_n; p_1, \ldots, p_n) \qquad (7)$$

*On introduction the variables $Z_d^{(s)} = e^i p_j^{(s)}$, $w_j^{(s)} = e^{iq_j^{(s)}}$, we can identify the manifold (6) with the manifold

$$\prod_{j\in\beta_i} Z_j^{(s)} = \prod_{j\in\beta_i'} w_j^{(s)}, \; |Z_j^{(s)}| = |w_j^{(s)}| = 1,$$

$$j = 1, \ldots, n, \quad s = 1, \ldots, v, \quad i = 1, \ldots, k$$

in $(C^v \times C^v)^n$. The fact that a_γ is analytic implies that it is analytic in a complex neighbourhood of this manifold.

where division γ^* is obtained from γ by permuting sets β_i and β_i' within every pair, the operator is self-adjoint.

Every cluster operator commutes with the group of operators U_s in L_n:

$$(U_s f)(p_1, \ldots, p_n) = \exp\{i(s, p_1 + p_2 + \ldots + p_n)f(p_1, \ldots, p_n\} \quad (8)$$

specifying the representation of translation group Z^ν in L_n.

We shall call the cluster operator symmetric if its kernel $K(p_1, \ldots, p_n, q_1, \ldots, q_n)$ is invariant under permutation of the variables $p_1, \ldots, p_n$ and the variables $q_1, \ldots, q_n$ separately. The subspaces $L_n^s = L_2^{sym}((T^\nu)^n)$ and $L_n^a = L_2^{asym}((T^\nu)^n)$ of, respectively, symmetric and anti-symmetric functions of the variables $p_1, \ldots, p_n$, $p_i \in T^\nu$, are invariant under the symmetric cluster operators; its parts, acting in these sub-spaces, will be called respectively *boson* and *fermion* cluster operators.

We will now show that any (unmarked) cluster operator, acting in the space $\ell_2(C^n_{Z^\nu})$ ($C^n_{Z^\nu}$ is the set of n-point subsets of Z^ν), transforms after conversion into a Boson cluster operator.

We define the Fourier transformation of the function $f \in \ell_2(C^n_{Z^\nu})$ as the mapping

$$F: \ell_2(C^n_{Z^\nu}) \to L^s_n: f \to \tilde{f}$$

$$= \sum_{T \in C^n_{Z^\nu}} f(T)\, e_T(p_1, \ldots, p_n) \quad (9)$$

where

$$e_T(p_1, \ldots, p_n) = \frac{1}{(n!)^{1/2}} \sum_\sigma \exp\left\{ i \sum_k (p_{\sigma(k)}, x_k) \right\}$$

$$T = \{x_1, \ldots, x_n\}$$

and the summation is over all premutations $\sigma: N_n \to N_n$ (the definition of e_T is independent of the numbering of points T). It is easily seen that this mapping is isometric; we denote the image of the space by $F\ell_2(C^n_{Z^\nu}) = \tilde{L}^s_n \subset L^s_n$. It is easily verified that its orthogonal complement $M_n = L^s_n \ominus \tilde{L}^s_n$ consists of functions of the form

$$\psi(p_1, \ldots, p_n) = \text{sym}\,(h(p_1 + p_2, p_3, \ldots, p_n)),$$

where sym denotes the operation of symmetrization with respect to all variables, and h is any function of L_{n-1}. Every operator A, acting in $\ell_2(C^n_{Z^\nu})$ with the aid of matrix $a_{TT'}$ (see (7.3)) transforms under mapping (9) into operator $\tilde{A}$, given by the kernel

$$K(p_1, \ldots, p_n; q_1, \ldots, q_n) = \sum_{T, T'} a_{T,T'} e_T(p_1, \ldots, p_n) e_T(q_1, \ldots, q_n) \tag{10}$$

We can regard the operator $\tilde{A}$ as acting in the entire space L_n, though it only acts nontrivially obviously in $\tilde{L}^s_n$, while the orthogonal complement $L_n \ominus \tilde{L}^s_n$ transforms to zero. In the case when A is a cluster operator in $\ell_2(C^n_{Z^\nu})$, i.e., $a_{T,T'}$ is written in the form (8.3), the right-hand side of (10) may be written as the sum of terms of the type (given suitable numbering of elements T and T′)

$$\omega_k[(T_1, T_1'), \ldots, (T_k, T_k')] \exp\left\{ i \left(\sum_k (x_{\sigma(k)}, p_k) - \sum_j (x'_{\sigma'(j)}, q_j) \right) \right\},$$

which, in view of (10.3), can be rewritten as

$$\begin{aligned} &\exp\{i(\hat{x}_1, (p_{\beta_1} - Q_{\beta_1'}))\} \ldots \exp\{i(\hat{x}_k, (p_{\beta_k} - Q_{\beta_k'}))\}. \\ &\cdot \prod_{m=1}^{k} \exp\left\{ i \left[\sum_{j\in\beta_m} (\tilde{x}_j, p_j) - \sum_{j\in\beta_m'} (\tilde{x}_j', q_j) \right] \right\} . \\ &\cdot \omega_k[(\tilde{T}_1, \tilde{T}_1'), \ldots, (\tilde{T}_k, \tilde{T}_k')] \end{aligned} \tag{11}$$

Here, $\hat{x}_m$ is the least point (in the sense of lexicographic ordering) T_m, $m = 1, \ldots, k$, $(\tilde{T}_m, \tilde{T}_m') = (T_m - \hat{x}_m, T_m' - \hat{x}_m)$ of the pair of sets, obtained by a shift of $-\hat{x}_m$ from (T_m, T_m'), while $\{(\beta_1, \beta_1'), \ldots, (\beta_k, \beta_k')\} = \gamma$ is a division of (N_n, N_n'), induced by the division of (T, T′) into pairs (T_m, T_m'). On then summing (11) with fixed division γ over points $\hat{x}_1, \ldots, \hat{x}_k$ and over sets of pairs $\{(\tilde{T}_1, \tilde{T}_1'), \ldots, (\tilde{T}_k, \tilde{T}_k')\}$ (in which all the first sets $\tilde{T}_m$ contain the point $0 \in Z^\nu$) in the light of estimate (6.3), we obtain the generalized function

$$\prod_{m=1}^{k} \delta(p_{\beta_m} - Q_{\beta_m'})\, a_\gamma(p_1, \ldots, p_n; q_1, \ldots, q_n), \tag{12}$$

where $a_\gamma(p_1, \ldots, p_n; q_1, \ldots, q_n)$ is an analytic function (regarded as a function of the variables $Z_j^{(s)} = \exp(i\, p_j^{(s)})$, $w_j^{(s)} = \exp(i\, q_j^{(s)})$, $j = 1, \ldots, n$, $s = 1, \ldots, \nu$) in the domain

$$1 - \epsilon < |Z_j^{(s)}|, |w_j^{(s)}| < 1 + \epsilon \tag{13}$$

where $\epsilon = \epsilon(\lambda, n) > 0$ is a constant, depending on the clustering parameter λ and the number n.

To sum up, a study of the spectral properties of the transfer matrix in its n-particle cluster subspaces reduces to a study of the spectrum of a self-adjoint Boson cluster operator, acting in L_n^s (more precisely, in $\tilde{L}_n^s \subset L_n^s$).

5.2. Algebra of Cluster Operators

We shall establish here some algebraic properties of the set K_n of cluster operators acting in L_n.

The most important is that K_n is an algebra of operators. The linear properties of K_n are obvious; in order to describe the multiplication rules in K_n, we introduce the following binary operation $\gamma_1 \bigcirc \gamma_2$ into the set of divisions (3) of the pair (N_n, N_n'). We write γ_1 as the division of the pair (N_n, N_n')

$$\gamma_1 = \{(\beta_1, \beta_1'), \ldots, (\beta_{k_1}, \beta_{k_1}')\}$$

$\beta_i \subseteq N_n$, $\beta_i' \subseteq N_n'$, while the division γ_2

$$\gamma_2 = \{(\widetilde{\beta_1'}, \beta_1''), \ldots, (\widetilde{\beta_{k_2}'}, \beta_{k_2}'')\}$$

is written as a division of the pair of sets (N_n', N_n'') (N_n'' is a further exemplar of the set $\{1, \ldots, n\}$).

For the two divisions

$$\alpha = (\beta_1', \ldots, \beta_{k_1}') \text{ and } \tilde{\alpha} = (\widetilde{\beta_1'}, \ldots, \widetilde{\beta_{k_2}'})$$

of the set N_n', we denote by

$$\epsilon = \alpha \vee \tilde{\alpha} = (\epsilon_1, \ldots, \epsilon_s)$$

the least division of N_n', whose blocks wholly include both the blocks β_i' and the blocks $\widetilde{\beta_j'}$ (ϵ is the upper bound of α and $\tilde{\alpha}$ in the structure of divisions of N_n', see [35]). Then, the division γ of the pair $\{N_n, N_n''\}$ into s pairs of subsets

$$\gamma = \{(\delta_1, \delta''_1), \ldots, (\delta_s, \delta_s'')\}$$

is "connected" with respect to ϵ, i.e., is such that

$$\delta_i = \bigcup_{\substack{m: \\ \beta_m' \subseteq \epsilon_i}} \beta_m, \quad \delta_i'' = \bigcup_{\substack{m: \\ \widetilde{\beta_m'} \subseteq \epsilon_i}} \beta_m'', \; i = 1, \ldots, s \tag{14}$$

which we denote in fact by $\gamma_1 \bigcirc \gamma_2$.

We now consider the two cluster operators $A\gamma_1$ and $A\gamma_2$ with "elementary" kernels of the type

$$K_{\gamma_i}(p_1, \ldots, p_n; q_1, \ldots, q_n) = a_{\gamma_i}(p_1, \ldots, p_n; q_1, \ldots, q_n)\,\delta_{\gamma_i} \qquad i = 1, 2 \tag{15}$$

Then the product $A_{\gamma_1} A_{\gamma_2}$ of two such operators is again specified by an elementary kernel

$$\tilde{a}_{\gamma_1 \circ \gamma_2}(p_1, \ldots, p_n, q_1, \ldots, q_n)\,\delta_{\gamma_1 \circ \gamma_2} \tag{15a}$$

where the function $\tilde{a}_{\gamma_1 \circ \gamma_2}$ is given by the integral

$$a_{\gamma_1 0 \gamma_2}(p_1, \ldots, p_n, q_i, \ldots, q_n) = \\ = \int_{\Gamma_{\gamma_1}^{[p_i]} \cap \Gamma_{\gamma_2}^{[q_i]}} a_{\gamma_1}(p_1, \ldots, p_n; s_1, \ldots, s_n)\, a_{\gamma_2}(s_1, \ldots, s_n; q_1, \ldots, q_n)\, ds_1 \ldots ds_n \tag{16}$$

taken over the intersection of manifolds*

$$\Gamma_{\gamma_1}^{[p_i]} = \{(s_1, \ldots, s_n): S_{\beta_i'} = p_{\beta_i}, i = 1, \ldots, k_1\}$$
$$\Gamma_{\gamma_2}^{[q_i]} = \{(s_1, \ldots, s_n): s_{\tilde{\beta}_i'} = Q_{\beta_i'}, i = 1, \ldots, k_2\}.$$

We call the number of pairs in division γ its rank $r(\gamma)$, while the rank $r(A)$ of cluster operator A is the maximum of the ranks $r(\gamma)$ of divisions γ, for which $a_\gamma \not\equiv 0$. It is easily seen from definition (14) that $r(\gamma_1 \circ \gamma_2) \leqslant \min(r(\gamma_1), r(\gamma_2))$, so that the set $H_n^{(s)} \subset K_n$ of cluster operators A, whose rank $r(A) \leqslant s < n$, forms a two-sided ideal in algebra K_n.

We shall mention a useful consequence of this fact. With each cluster operator A we associate a chain of cluster operators

$$A^{(n)}, A^{(n-1)}, \ldots, A^{(1)} = A \tag{17}$$

in such a way that the kernel of the operator $A^{(s)}$ is equal to the sum of the elementary terms $a_\gamma \delta_\gamma$ in expansion (2) for the kernel of A, the rank of which is not less than s. Obviously, for the product $A_1 . A_2$ of two cluster operators, the s-th element of chain (17) is determined solely by the m-th elements of the chains for A_1 and A_2 with $m \geqslant s$

$$(A_1 A_2)^{(s)} = \sum_{m \geqslant s} (A_1^{(m)} A_2^{(m)})^{(s)} \tag{18}$$

*It is easily verified that the manifold $\Gamma_{\gamma_1}^{[p_i]} \cap \Gamma_{\gamma_2}^{[q_i]}$ is not empty and its dimensionality is equal to the number of independent cycles in the following graph: its vertices are the lower blocks $\{\beta_1', \ldots, \beta_{k_1}'\}$ in γ_1 and the upper blocks $\{\tilde{\beta}_1', \ldots, \tilde{\beta}_{k_2}'\}$ in γ_2, while its ribs are the pairs $(\beta_i', \tilde{\beta}_j')$, for which $\beta_i' \cap \tilde{\beta}_j' \neq \phi$; (in the case when there are no such cycles, i.e., $\Gamma_{\gamma_1}^{[p_i]} \cap \Gamma_{\gamma_2}^{[q_i]}$ consists of a single point, the integral (16) is replaced by the value of the integrand at this point).

5.3. Spectrum and Resolvent of the Finite-Particle Cluster Operator

We shall describe here the spectrum of the transfer matrix T in its k-particle invariant subspaces H_k.

(1) We start with the case $k=1$.
For the simplest (unmarked) one-particle (cluster) invariant subspace H_1, in which the operator $T_1 = T|_{H_1}$ is unitarily equivalent to the convolution operator with function a_s (see 1.4), acting in $\ell_2(Z^v)$, we find after passage to the Fourier transformation that T_1 and group U_s in H_1 are unitarily equivalent to the operators in $L_2(T^v)$:

$$(\tilde{T}_1 f)(p) = \epsilon(p)f(p), \; (U_s f)(p) = e^{i(s,p)}f(p) \tag{19}$$

where $\epsilon(p) = \sum_s a_s e^{i(p,s)}$ is the Fourier transformation of the function a_s. Hence the spectrum of the operator T_1 in H_1 is Lebesgue and is the same as the set of values of the function $\epsilon(p)$ (the one-particle branch of the spectrum of the transfer matrix T).

In the more general case of a marked one-particle subspace H_1 (see Para. I. 4), we find that the operator T_1 in H_1 is unitarily equivalent to the operator

$$(\tilde{T}_1 f)_\gamma (p) = \sum_{\gamma' \in M} B_{\gamma,\gamma'} (p) f_{\gamma'}(p), \; \underset{\gamma \in M}{p \in T^v} \tag{20}$$

acting in the space $L_2(T^v) \otimes \ell_2(M)$ of the functions $\{f_\gamma(p), p\in T^v, \gamma \in M\}$, (M is the set of indices introduced in Section 2.2), $\{B_{\gamma,\gamma'}(p)\}$ is a self-adjoint matrix, specifying, for any p, a bounded operator B(p) in $\ell_2(M)$ (see estimates (3.4)). If we assume that M is finite and denote by $\epsilon_1(p)$, . . ., $\epsilon_k(p)$ the eigenvalues of B(p) (numbered in decreasing order of absolute value), then the spectrum of T_1 is the same as the union $\cup$ Im $\epsilon_1(p)$ of the ranges of values of the functions ϵ_i, $i=1, \ldots, k$, and is purely Lebesgue. Notice that, if, with $p \in T^v$, the values of a branch $\epsilon_{i_0}(p)$ are different from the values of the other branches, then $\epsilon_{i_0}(p)$ is a smooth (analytic) function of p, and we can isolate in H_1 an (unmarked) cluster one-particle subspace, invariant with respect to T_1 and U_s, in which the spectrum of T_1 is the branch ϵ_{i_0}(see (19)).

In the case when, for any $p \in T^v$, the values of $\epsilon_i(p)$ are the same as the values of the other branches, the function $\epsilon_i(p)$ is not in general smooth, and there cannot be a cluster invariant subspace $H_1^i \subset H_1$ with a branch of the spectrum $\epsilon_i(p)$.

(2) Case $k=2$. We consider for simplicity the case of a two-particle unmarked cluster subspace H_2 of the transfer matrix, in which, as we know, its part $T_2 = T|_{H_2}$ and group U_s are unitarily equivalent to operators (4.4) and (5.4) respectively, acting in $\ell_2(C^{(2)}{}_{Z^\nu})$; after a Fourier transformation these operators become the operators

$$(\tilde{T}_2 f)(p_1, p_2) = a_1(p_1, p_2)f(p_1, p_2) + \\ + \int\limits_{p_1+p_2=q_1+q_2} a_2(p_1, p_2; q_1, q_2)f(q_1, q_2)\, dq_1\, dq_2 \tag{21}$$

and

$$(U_s f)(p_1, p_2) = \exp\{i(s, p_1+p_2)\}\, f(p_1, p_2),$$

acting in the space $\tilde{L}_2^s$ of symmetric functions $f(p_1, p_2)$ such that

$$\int\limits_{p_1+p_2=p} f(p_1, p_2)\, dp_1\, dp_2 = 0$$

for any $p \in T^\nu$.

Here, a_1 and a_2 are the cluster functions obtained from cluster functions ω_1 and ω_2 in (5.4).

We next isolate the "total momentum," i.e., we introduce variables $P = p_1 + p_2$, $p = p_1$ and write the space $\tilde{L}_2^s$ as the direct integral

$$\tilde{L}_2^s = \int \tilde{L}_2(p)\, dP,$$

where $\tilde{L}_2(P) \subset L_2(T^\nu)$ is the space of functions $f \in L_2(T^\nu)$ such that

$$f(p) = f(P-p), \int\limits_{T^\nu} f(p)dp = 0 \tag{21^a}$$

It can be seen from (21) that the operators $\tilde{T}_2$ and U_s here can also be expanded in direct operator integrals

$$\tilde{T}_2 = \int \tilde{T}_2(P)dp,\ U_s = \int U_s(P)dp,$$

where $\tilde{T}_2(p)$ and $U_s(p)$ act in $\tilde{L}_2(P)$:

$$(\tilde{T}_2(P)f)(p) = \epsilon_p(p)f(p) + \int\limits_{T^\nu} k_P(p, q)f(q)dq$$
$$\epsilon_p(p) = a_1(p, P-p),\ K_p(p, q) = a_2(p, P-p, q, P-q)$$
$$(U_s(p)f)(p) = e^{i(s,p)}f(p) \tag{22}$$

The class of operators (22) – the so-called "Friedrichs model" – has often been studied (the references can be found in [43]). On using these results, it can be shown that there exists in the space $\tilde{L}_2(P)$ a subspace

$\tilde{L}_2^0(P) \leqslant \tilde{L}_2(P)$, invariant under the operator $\tilde{T}_2(P)$, in which it is unitarily equivalent to the operator of multiplication by the function $\epsilon_p(.)$ in $L_2(T^\nu)$ (the Lebesgue branch of the spectrum), and possibly to a finite number of eigenvectors with eigenvalues $\epsilon^{(1)}(p), \ldots, \epsilon^{(k)}(p)$, numbered in increasing order and located outside the continuous spectrum of $\tilde{T}_2(p)$ (i.e., outside the set of values of the function ϵ_p; with some exceptional values of p there may appear eigenvalues $\tilde{T}_2(p)$, lying in the continuous spectrum; however, they make no contribution to the spectrum of $\tilde{T}_2$ and can be ignored). Each of eigenvalues $\epsilon^{(i)}(p)$ is a continuous function of p, defined in a domain $G_i \subseteq T^\nu$. Thus, on passing to the entire operator $\tilde{T}_2$ in $\tilde{L}_2^s$, we see that there is a subspace $L_1^0 \subset \tilde{L}_2^s$, invariant with respect to $\tilde{T}_2$ and U_s, in which the operator $\tilde{T}_2$ is unitarily equivalent to the operator of multiplication by the function $a(p_1, p_2) = \epsilon_{p_1+p_2}(p_1)$, acting in L_2^s, and the operator U_s is unitarily equivalent to the operator (21) (the two-particle branch of the spectrum). In addition, there are possibly invariant subspaces $L_1, \ldots, L_k$ - so-called subspaces of "bound states of two particles," - in each of which $\tilde{T}_2$ is unitarily equivalent to the operator of multiplication by the function $\epsilon^{(i)}(p)$, acting in the space $L_2(G_i)$ of functions f(p), defined in the domain G_i (the operator U_s is then unitarily equivalent in each L_i to multiplication by $\exp\{i(s, P)\}$).

Mamatov and Minlos [24] (see also Mamatov [25]) studied in detail the case of operator (21), typical for the transfer matrices of Gibbs spin fields with small β and for dimensionalities $\nu = 1, 2, 3$. In this case, $a_1(p_1, p_2)$ has the form

$$a_1(p_1, p_2) = \beta^2 a_0(p_1, p_2) + \tilde{a}_1(p_1, p_2),$$

where $a_0(p_1, p_2)$ is a function which is independent of β and differs from zero, while the function $\tilde{a}_1(p_1, p_2)$, and also $a_2(p_1, p_2; q_1, q_2)$, are of order β^3. It is assumed here that the family of functions $\{\epsilon_p^{(0)}(p) = a_0(p, P-p), P \in T^\nu\}$ is a family of general position: given any $P \in T^\nu$, the function $\epsilon_p^{(0)}$ has only a finite number of critical points; the caustic set of values P (i.e., of the P, for which one critical point of $\epsilon_p^{(0)}$ degenerates) consists of elementary caustics (see [48]), while the type of singularity at degenerate points is the minimum possible for the given dimensionality ν of the parameter P (again, see [48]); the set of "multiple values" P, at which there are multiple critical values, then consists of smooth $(\nu-1)$-dimensional manifolds. The analysis in [24] shows that, in the case described, with $v=3$, the operator $\tilde{T}_2$ has no bound states, while with $v=1$ or 2, they can appear only in a small neighborhood of the momentum P values at which there are multiple extremal values of $\epsilon_P^{(0)}$. (3) While study of the spectrum of the transfer matrix in its three-particle cluster invariant subspace has been undertaken by Mamatov and Minlos in [23], and by Minlos and Mogilner in [22], it has not yet reached the same degree of completeness as in the two-particle case. This problem is similar in its ideas to the problem of the spectrum and scattering of the three-particle Schrödinger operator and may be solved by similar devices (see D.D. Faddeyev [49], and Merkur'ev and Faddeyev [50]).

To sum up, we know that the part of the transfer matrix $T_3 = T|_{H_3}$ in its (unmarked) three-particle cluster invariant subspace H_3 is unitarily equivalent to the boson cluster operator $\tilde{T}_3$, acting in space $\tilde{L}_3^s$ and given by the series

$$
\begin{aligned}
K(p_1, p_2, p_3, q_1, q_2, q_3) &= a_1(p_1, p_2, p_3). \\
&\prod_{i=1}^{3} \delta(p_i - q_i) + a_2(p_1, p_2, p_3; q_1, q_2, q_3). \qquad (23) \\
&\delta(p_1 - q_1)\,\delta(p_2 + p_3 - q_2 - q_3) + \hat{a}_2(p_1, p_2, p_3: \\
&q_1, q_2, q_3)\delta(p_1 - q_2 - q_3)\delta(p_2 + p_3 - q_1) +
\end{aligned}
$$

+ terms obtained by permuting variables p_1, p_2, p_3 and q_1, q_2, q_3 + $a_3(p_1, p_2, p_3, q_1, q_2, q_3)\,\delta(p_1 + p_2 + p_3 - q_1 - q_2 - q_3)$, where the cluster functions $a_2(p_1, p_2, p_3; q_1, q_2, q_3)$, $\hat{a}_2(p_1, p_2, p_3; q_1, q_2, q_3)$ are defined respectively on manifolds $\Gamma_2 = \{p_1 = q_1,\ p_2 + p_3 = q_2 + q_3\}$ and $\hat{\Gamma}_2 = \{p_1 = q_2 + q_3,\ p_2 + p_3 = q_1\}$, while function $a_1(p_1, p_2, p_3; q_1, q_2, q_3)$ is defined on manifold $\Gamma_3 = \{p_1 + p_2 + p_3 = q_1 + q_2 + q_3\}$.

We now assume that, for all regular values Z, the resolvent $R(Z) = (\tilde{T}_3 - ZE)^{-1}$ is also a cluster operator, i.e., its kernel can be written in the form (23) with the cluster functions

$$
\begin{gathered}
r_1(p_1, p_2, p_3; Z),\ r_2(p_1, p_2, p_3; q_1, q_2, q_3; Z),\ \hat{r}_2(p_1, p_2, p_3; \\
q_1, q_2, q_3; Z),\ r_3(p_1, p_2, p_3; q_1, q_2, q_3; Z).
\end{gathered}
$$

Using the identity

$$(\tilde{T}_3 - ZE)\,R(Z) = E,$$

along with the rule for multiplication of cluster operators (15[a]) and (16),

Note (18), and the fact that E is also a cluster operator with cluster functions $a_1 \equiv 1$, $a_2 = \hat{a}_2 = a_3 \equiv 0$, we obtain the following relations for the cluster functions

$$1. \qquad (a_1(p_1, p_2, p_3) - Z)\, r_i(p_1, p_2, p_3) = 1 \tag{24}$$

$$\begin{aligned}2. \qquad & (a_1(p_1, p_2, p_3) - Z)\, r_2(p_1, p_2, p_3; q_1, q_2, q_3) \\ & + 3 \int_{\substack{s_1 = p_1 = q_1 \\ s_2 + s_3 = p_2 + p_3 = q_2 + q_3}} a_2(p_1, p_2, p_3; s_1, s_2, s_3)\, r_2(s_1, s_2, s_3; q_1, q_2, q_3)\, ds_1\, ds_2\, ds_3 \\ & + 3 \int_{\substack{s_1 = p_2 + p_3 = q_2 + q_3 \\ s_2 + s_3 = p_1 = q_1}} \hat{a}_2(p_1, p_2, p_3; s_1, s_2, s_3)\, \hat{r}_2(s_1, s_2, s_3; q_1, q_2, q_3)\, ds_1\, ds_2\, ds_3 + \delta_1 = 0\end{aligned} \tag{25}$$

$$\begin{aligned} & (a_1(p_1, p_2, p_3) - Z)\, \hat{r}_2(p_1, p_2, p_3; q_1, q_2, q_3) \\ & + 3 \int_{\substack{s_1 = p_1 = q_2 + q_3 \\ s_2 + s_3 = p_2 + p_3 = q_1}} a_2(p_1, p_2, p_3; s_1, s_2, s_3)\, \hat{r}_2(s_1, s_2, s_3; q_1, q_2, q_3)\, ds_1\, ds_2\, ds_3 \\ & + 3 \int_{\substack{s_2 + s_3 = p_1 = q_2 + q_3 \\ s_1 = p_2 + p_3 = q_1}} \hat{a}_2(p_1, p_2, p_3; s_1, s_2, s_3)\, r_2(s_1, s_2, s_3; q_1, q_2, q_3)\, ds_1\, ds_2\, ds_3 + \hat{\delta}_2 = 0\end{aligned}$$

$$3. \qquad (\widetilde{T}_3 - ZE) r_3 + \delta_3 = 0 \tag{26}$$

Here, δ_2 and $\hat{\delta}_2$ are functions, explicitly expressible in terms of r_1, while δ_3 is explicitly expressible in terms of r_2 and $\hat{r}_2$.

The above relations can be regarded as a system of equations in the cluster functions $r_1, r_2, \hat{r}_2, r_3$. Notice that this system, as is to be expected from Note 18, is of a recurrence type: from the first Eq. (24) we find that

$$r_1(p_1, p_2, p_3) = \frac{1}{a_1(p_1, p_2, p_3) - Z} \tag{27}$$

provided that $Z \overline{\in} \phi_1 \equiv \operatorname{Im} a(p_1, p_2, p_3)$.

Next, under the same condition, the second two equations (25) in r_2, $\hat{r}_2$ can be rewritten in the form

$$(A_{P_1, P_2} - ZE)\varrho_{P_1, P_2, q} = -\delta_{P_1, P_2, q} \tag{28}$$

where $P_1 = p_1$, $P_2 = p_2 + p_3$, A_{P_1, P_2} is a family of operators, acting respectively in spaces $\widetilde{L}_2(P_2) \oplus \widetilde{L}_2(P_2)$ of pairs of functions $\varrho = (\varrho_1(p), \varrho_2(p))$, satisfying conditions (21[a]) (with $P = P_2$),

$$(A_{P_1,P_2}\varrho)_i(p) = \epsilon_{P_1,P_2}(p)\varrho_i(p) + \int_{T^v} \Big(\sum_{j=1,2} K^{(i,j)}_{P_1,P_2}(p, s). \varrho_j(s)\Big)\, ds,$$

where $\epsilon_{P_1,P_2}(p) = a_1(P_1, p, P_2 - p)$, while kernels $K^{(i,j)}_{P_1,P_2}$ are given in terms of the functions a_2 and $\hat{a}_2$:

$$K^{(1,1)}_{P_1,P_2}(P, s) = 3\, a_2(P_1, p, P_2 - p; P_1, s, P_2 - s),$$

etc., while

$$\varrho^{(p)}_{P_1,P_2,q} = (r_2(P_1, p, P_2 - p; P_1, q, P_2 - q), \hat{r}_2(P_1, p, P_2 - p; P_2, q, P_1 - q))$$

and $\delta_{P_1,P_2,q}$ is similarly expressible in terms of δ_2 and $\hat{\delta}_2$.

The operator A_{P_1,P_2} is similar to the operator (22) (Friedrichs model), and in addition to the continuous spectrum $\mathrm{Im}\,\epsilon_{P_1,P_2} \subset \phi_1$, possibly has a finite number of eigenvalues $\epsilon^{(i)}_{(P_1,P_2)}$. The functions $\epsilon^{(i)}_{(P_1,P_2)}$ (branches of the spectrum of two-particle bound states) are each defined in a domain $G^{(i)} \subseteq T^v \times T^v$ and are continuous. We put $\phi_2 = \bigcup_i \mathrm{Im}\,\epsilon^{(i)}$; in the case when $Z \bar{\in} \phi_2 \cup \phi_1$, Eq. (28) is solvable for all P_1, P_2, q and defines the cluster functions r_2 and $\hat{r}_2$. We turn to the last equation (26). We construct the cluster operator $\hat{R}(Z)$ with cluster functions r_1, r_2, $\hat{r}_2$, and $r_3 \equiv 0$ (the operator $R^{(2)}(Z)$ in the notation (17)). By (18),

$$\begin{aligned} \hat{R}(Z)\,(\widetilde{T}_3 - ZE) &= E + N_{lft}(Z) \\ (\widetilde{T}_3 - ZE)\,\hat{R}(Z) &= E + N_{rgt}Z \end{aligned} \qquad (29)$$

where N_{lft} and N_{rgt} are cluster operators belonging to the ideal $H^{(1)}_3$ (i.e., such that their cluster functions $a_1 = a_2 = \hat{a}_2 = 0$).

Further, if we consider for each fixed value $P = p_1 + p_2 + p_3$ of the total momentum the restrictions $\widetilde{T}_3(P)$, $\hat{R}(P; Z)$, $N_{lft,rgt}(P)$ of operators $\widetilde{T}_3$, $\hat{R}(Z)$, and $N_{lft,rgt}$ on to the subspaces $\widetilde{L}^s_3(P)$ of symmetric functions, given on the manifold $\{(p_1, p_2, p_3): p_1 + p_2 + p_3 = P\}$, then Eq. (26) can be rewritten as

$$(\widetilde{T}_3(P) - ZE)\varrho_{P,q_1,q_2} = -\delta_{P,q_1 q_2} \qquad (30)$$

where $\varrho^{(p_1,p_2)}_{P,q_1 q_2} = r_3(p_1, p_2, P - p_1 - p_2; q_1 q_2, P - q_1 - q_2)$ and the expression for $\delta_{P,q_1 q_2}$ is similar.

Since, for any P, the operators $N_{lft,rgt}(P)$ are compact, relations (29), rewritten for $\widetilde{T}_3(P)$; $\hat{R}(P; Z)$ and $N_{lft,rgt}(P)$, mean that the operator $(\widetilde{T}_3(P) - ZE)$ with $Z \bar{\in} \phi_1 \cup \phi_2$ is Fredholm (with zero index, since it is self-adjoint), and hence there is only a denumerable (and in practice, finite, for the typical cases) number of eigenvalues $\epsilon_j(P)$, $j = 1, 2, \ldots,$

of operator $\tilde{T}_3(P)$. The functions $\epsilon_j(P)$ (branches of three-particle "bound states"), which are each defined in its own domain $G_j \subset T^\nu$, are continuous; we denote by ϕ_3 the set $\phi_3 = \bigcup_j \operatorname{Im} \epsilon_j$. Then, if $Z \bar{\in} \phi_1 \cup \phi_2 \cup \phi_3$, a solution of Eq. (30) exists for any P and defines the cluster function r_3 of the resolvent R(Z).

A more detailed analysis, using the methods of scattering theory, shows that the set $\phi_1 \cup \phi_2 \cup \phi_3$ is the same as the spectrum of $\tilde{T}_3$, and the space $\tilde{L}_3^s$ splits into the orthogonal sum of subspaces $L_3^{(0)}$, $L_i^{(1)}$, $L_j^{(2)}$, invariant with respect to $\tilde{T}_3$ and U_s, and in each of them the operator $\tilde{T}_3$ is unitarily equivalent to the operator of multiplication by the respective branch $\epsilon(p_1\, p_2\, p_3) = a_1(p_1\, p_2\, p_3)$, $\epsilon^{(i)}(P_1, P_2)$, $\epsilon_j(P)$, while the operator U_s has the form (8) (or respectively, (21) or (19)).

The device described can also be used to find the spectrum in the 4-, 5-, etc. k-particle invariant subspaces and thus to isolate the total "unbound" branch of the spectrum, and the branches of the two-, three-, four-particle etc. bound states in spaces H_k. However, this program involves certain analytic difficulties which have not yet been overcome, and are linked to describing the behavior of the resolvent in the neighbourhood of "cuts" of the spectrum.

Note. The scheme described here for studying the three-particle resolvent of the cluster operator is similar to the scheme proposed by Weinberg [52] and Van Vinter [51] for studying the three-particle Schrödinger operator, while the equation

$$(E + N_{lft})r_3 = -\hat{R}(Z)\delta,$$

which follows from (29) and (30), is a generalization of the Weinberg-Van Vinter equation for the bound part of the resolvent of this operator (see M. Reed and B. Simon [43]).

6. Supplementary Topics and Applications

We shall briefly mention here some topics relevant to the above material.

6.1. Decrease of Correlation of the Gibbs Field

Recent studies of the structure of the spectrum and resolvent of the transfer matrix in its leading (one- and two-particle) invariant subspaces can also be utilized to find the asymptotic behavior, with $\beta \ll 1$, of the decrease of the correlation of functionals which depend on the field values at points remote from one another.

We consider in the simplest case the correlation $s_1(t) = <\sigma_{(0,0)}, \sigma_{(0,t)}> = <\sigma_{(0,0)} \cdot \sigma_{(0,t)}> - <\sigma_{(0,0)}>^2$ of the values of a spin field at points $(0, 0), (0, t) \in Z^{\nu} \times Z^1$. Using the expansion

$$\sigma_{(0,0)} - <\sigma_{(0,0)}> = h_1 + h_2,$$

where $h_1 \in H_1$, $h_2 \in H_1^{\perp}$ (H_1 is the one-particle subspace), we find that

$$\begin{aligned} s_1(t) &= ((\sigma_{(0,0)} - <\sigma_{0,0}>, T^t(\sigma_{(0,0)} - <\sigma_{(0,0)}>)) \\ &= (h_1, T_1^t h_1) + (T^t h_2, h_2) \end{aligned} \tag{1}$$

On then passing to the spectral representation for T_1 (the p-representation), we see that

$$(h_1\ T_1^t\ h_1) = \int_{T^{\nu}} \epsilon^t(p)\ |\varphi(p)|^2\ dp \tag{2}$$

where $\epsilon(p)$ is the one-particle branch of the spectrum T in H_1, and φ is the Fourier transform of the vector $h_1 \in H_1$. From (2), provided that the maximum point $p_0 \in T^{\nu}$ of the function $\epsilon(p)$ is single and nondegenerate, and $\varphi(p_0) \neq 0$, we find that

$$(h_1, T_1^t\ h_1) = \frac{C}{t^{\nu/2}} (\epsilon(p_0))^t\ (1 + 0(1)) \text{ as } t \to \infty \tag{3}$$

Since $(T^t\ h_1, h_2) \sim (C\beta)^{2t}$, the left-hand side of (3) in fact gives the asymptotic form of $s_1(t)$. We can evaluate in a similar way the asymptotic form of correlation $<F_A, F_{A+t}>$, where $A \subset Y_0$, and $F_A \in H$ is a functional of the field values in the set A, which has a nonzero projection onto the subspace H_1. However, in the case of an interaction potential which is even (with respect to the variable σ_x), the product σ_{x_1} σ_{x_2}, $x_1, x_2 \in Y_0$, is orthogonal to H_1, and hence the correlation

$$s_2(t) = <\sigma_{x_1} \sigma_{x_2}, \sigma_{x_1 + te_{\nu+1}} \cdot \sigma_{x_2 + te_{\nu+1}}> = (h_2, T_2^t h_2) + O(\beta^{3t}) \quad t \in Z^1 \tag{4}$$

where T_2 is the part of the transfer matrix in a two-particle subspace H_2, and h_2 is the projection of $\sigma_{x_1} \sigma_{x_2}$ onto this subspace. Thus the asymptotic form of $s_2(t)$ as $t \to \infty$ is found from the spectrum of T_2 in H_2, and is equal to

$$s_2(t) \sim C\,\frac{(\epsilon(p_1^0, p_2^0))^t}{t^v} \text{ for } v \geqslant 3 \qquad (5)$$

In the case of dimensionality $v = 1$ the factor in front of the exponential is $1/t^2$, or which $v = 2$, it is $1/t^2 \ell n\, t$.

For more details about this, see Minlos and Sinai [31], Polyakov [45], and Minlos [32].

6.2. The Bethe–Salpeter Kernel

We mentioned in the Introduction that there is a somewhat different method of finding the two-, three-, . . ., k-particle bound states of a transfer matrix, which uses calculations with so-called Bethe-Salpeter kernels (see Glimm and Jaffe [36], Spenser [37], Spenser and Zirrili [38], Dimock and Eckmann [39], Shor [40], and O'Carroll [41].

We will briefly consider this method as applied to the transfer matrix of a spin Gibbs field, defined by even interaction potentials of type (16.1).

We consider the function

$$s_1(x, y) = \langle \sigma_x, \sigma_y \rangle = \langle \sigma_x \cdot \sigma_y \rangle$$
$$x = (\hat{x}, x_{v+1}) \in Z^v \times Z^1,\ y = (\hat{y}, y_{v+1}) \in Z^v \times Z^1.$$

Obviously,

$$s_1(x, y) = s_1(x - y) = (\sigma_{0,0}, U_{\hat{x}-\hat{y}}\, T_{\sigma_{0,0}}^{|y_{v+1} - x_{v+1}|})$$
$$= (U_{\hat{x}-\hat{y}} T_1^{|x_{v+1} - y_{v+1}|} h_1, h_1) + (U_{\hat{x}-\hat{y}} \cdot T^{|x_{v+1} - y_{v+1}|} h_2, h_2) \qquad (6)$$

where h_1 and h_2 are the projections of $\sigma_{0,0}$ onto H_1 and $H_1^\perp$ respectively.

Using the results of the previous chapter (see also the previous section), we find that

$$(U_{\hat{\xi}}\, T_1{}^{|\xi_0|} h_1, h_1) = \int_{T^v} e^{i(p,\hat{\xi})} (\epsilon(p))^{|\xi_0|}\, |\varphi(p)|^2\, dp$$

$$(U_{\hat{\xi}}\, T^{|\xi_0|} h_2, h_2) \sim 0\,(\beta^{2|\xi_0|})\ \hat{\xi} \in Z^v,\ \xi_0 \in Z^1. \qquad (7)$$

where

$$\epsilon(p) > C\,\beta \qquad (8)$$

and $C > 0$ is a constant.

Hence the Fourier transform of $s_1\,(\hat{\xi}, \xi_0)$ is

$$\tilde{s}_1(p, k) = \sum_{\hat{\xi}, \xi_0} s_1(\hat{\xi}, \xi_0)\, e^{-i(p, \hat{\xi})}\, e^{-ik\xi_0}$$

and is equal to

$$\tilde{s}_1(p, k) = |\varphi(p)|^2 \sum_{\xi_0 \in Z^1} (\epsilon(p))^{|\xi_0|}\, e^{-ik\xi_0} + \tilde{s}_1'(p, k) \tag{9}$$

where, by (7), $\tilde{s}_1'(p, k)$ is, for any fixed $p \in T^v$, an analytic function of the variable k in the strip

$$|\mathrm{Im}\, k| < 2|\ell n\, \beta| \tag{10}$$

The first sum in (9) is a meromorphic function of the variable k and in the strip (10) it has two poles at the points $k_0 = \pm i\, \ell n\, \epsilon(p)$.

Thus the spectrum of the transfer matrix in the one-particle subspace H_1 can be found from the poles of the function $\tilde{s}_1(p, k)$ as a function of the complex variable k ($p \in T^v$ is fixed).

To find the two-particle bound state of the transfer matrix in its invariant subspace H_2, we introduce the function

$$\begin{aligned} s(T, T') = {} & <(\sigma_{x_1} \sigma_{x_2} - <\sigma_{x_1} . \sigma_{x_2}>) . (\sigma_{y_1} . \sigma_{y_2} - \\ & <\sigma_{y_1} . \sigma_{y_2}>)> = <\sigma_{x_1} . \sigma_{x_2} . \sigma_{y_1} . \sigma_{y_2}> - <\sigma_{x_1} . \sigma_{x_2}> . \\ & <\sigma_{y_1} . \sigma_{y_2}>. \end{aligned} \tag{11}$$

where $T = (x_1, x_2) \in Z^{v+1}$, $T' = (y_1, y_2) \in Z^{v+1}$ are two-point subsets of Z^{v+1}, and we use it to define the operator

$$(Rf)(T) = \sum_{T'} S(T, T')f(T'),$$

acting in $\ell_2(C^2_{Z^{v+1}})$. We use the following expansion, which follows from the expressions for the moments in terms of the semi-invariants (see [35]):

$$\begin{aligned} S(T, T) = {} & s_1(x_1 - y_1)\, s_1(x_2 - y_2) + s_1(x_1 - y_2). \\ & s_1(x_2 - y_1) + S^C_y(T, T') \end{aligned} \tag{12}$$

where $s_1(x - y) = <\sigma_x, \sigma_y> = <\sigma_x . \sigma_y>$ (since $<\sigma_x> = 0$), and $S^C(T, T')$ is the semi-invariant of the field values at points (x_1, x_2, y_1, y_2), and we also use the cluster estimates of the semi-invariants; it then follows that R is a cluster operator (connected with (5.4)).

We introduce another cluster operator R_0, acting by means of the kernel

$$S_0(T, T') = s_1(x_1 - y_1)\, s(x_2 - y_2) + s_1(x_1 - y_2)\, s_1(x_2 - y_1) \tag{13}$$

By (8), $R_0 > C\beta^2 E$ ($C > 0$ is a constant), and hence the operators R and R_0 are invertible, and their inverses R^{-1} and R_0^{-1} are likewise cluster operators, i.e., the matrix elements $(R^{-1})_{T,T'}$ and $(R_0^{-1})_{T,T'}$ admit of expansion of type (5.4) with cluster functions ω_1, ω_2 and ω_1^0, ω_2^0 respectively. Here,

$$\omega_1 = \omega_1^0,$$

so that

$$(R^{-1})_{T,T'} - (R_0^{-1})_{T,T'} = \omega_2(T, T') - \omega_2^0(T, T') \overset{\mathrm{def}}{=} K(T, T') \tag{14}$$

We call K with matrix elements (kernel) $K(T, T')$ the Bethe–Salpeter operator. From definition (14) we see that

$$R + RKR_0 = R_0 \tag{15}$$

Passing to the p-representation (in space $L_2^{sym}(T^{v+1} \times T^{v+1})$) and separating the total $(v+1)$-dimensional momentum (p, k), $p \in T^v$, $k \in T^1$, we obtain the family of operators $R_0(p, k)$, $R(p, k)$, $K(p, k)$, connected by relation (15). Hence

$$R(p, k) = (E + K(p, k)R_0(p, k))^{-1} R_0(p, k) \tag{16}$$

Using the previous estimates for $\tilde{s}_1(p, k)$ and the cluster estimates for the semi-invariant $S_4^c(T, T')$, it can be shown that the operators $R_0(p, k)$ and $K(p, k)$ are analytic with respect to k for each fixed p in the strip

$$|\operatorname{Im} K| < \min_{p_1 + p_2 = P} |\ell n\, \epsilon(p_1) + \ell n\, \epsilon(p_2)| \tag{17}$$

We denote by $h_t^\varphi(\sigma)$ the functional of the field σ:

$$h_t^\varphi(\sigma) = \sum_{T \subset Y_t} \varphi(T)(\sigma_{x_1} \cdot \sigma_{x_2} - <\sigma_{x_1} \cdot \sigma_{x_2}>), \quad T = \{x_1, x_2\}$$

where $\varphi(T) \in \ell_2(C_{Z^v}^{(2)})$.

Obviously,

$$<h_0^\varphi \cdot h_t^\varphi> = \sum_{T \in Y_0,\, T' \subset Y_1} s(T, T')\, \varphi(T) \cdot \varphi(T') = (h_0^\varphi, T^t h_0^\varphi) \tag{18}$$

Since the vector $h_0^\xi \in H$ is orthogonal to constants and to the one-particle subspace H_1 (since the potential is even), it can be written as

$$h_0^\xi = \hat{h}_0^\xi + g^\varphi, \ \hat{h}_0^\xi \in H_2, \ g^\varphi \in H_2^\perp.$$

Using the results of Section 4, we find that

$$(h_0^\xi, T^t h_0^\xi) = (\hat{h}_0^\xi, T_2^t \hat{h}_0^\xi) + 0(\beta^{3t}).$$

On passing to the p-representation in (18), we find that, for any fixed p,

$$(h_0^\xi, T^t(p) h_0^\xi) = \int \tilde{S}(P, k; p_1, s_1, p_2, s_2)$$
$$\tilde{\varphi}(P, p_1) \tilde{\varphi}(p, p_2) e^{ikt} dp_1 dp_2 ds_1 ds_2 dk,$$

where $\tilde{s}$ is the Fourier transform of the kernel s.

In the case when the operator $T_2(P)$ in $H_2(P)$ has an eigenvalue $\epsilon_1(P)$, lying to the right of its continuous spectrum:

$$\epsilon_1(P) > \max_{p_1 + p_2 = P} \epsilon(p_1) \epsilon(p_2)$$

with eigenfunction $\varphi_0(P, p)$, then, on choosing it as φ, we find that

$$\sum_t C(P) \epsilon_1^t(P) e^{-ikt} = \int \tilde{S} (P, k, p_1, s_1,$$
$$p_2, s_2) \varphi_0(P, p_1) \varphi_0(P, p_2) dp_1 dp_2 ds_1 ds_2$$
$$= (R(P, k) \varphi_0(P, .), \varphi_0(P, .))$$

(C(P) is a constant).

Hence it is clear that R(P, k) has poles at the points

$$K_1 = \pm i \ \ell n \ \epsilon_1(P),$$

lying inside the strip (17).

On the other hand, it can be seen from (16) and (17) that, with this value of k, the compact operator $K(P, k)R_0(P, k)$ has an eigenvalue equal to -1. Hence the spectrum of the bound two-particle state with fixed $p \in T^\nu$, can be found as the zero of the function of the complex variable k:

$$D_P(k) = \text{Det} (E + K(P, k)R_0(P, k)), \tag{19}$$

inside the strip (17).

6.3. Other Examples of Cluster Operators

Notice that, in addition to the transfer matrices of Gibbs random fields, the transfer matrices of Markov random fields with local interaction (see e.g., [52]) are also cluster operators (in the case when the interaction is sufficiently small).

Another vast stock of cluster (additive) operators is provided by the Hamiltonians of quantum spin systems, arising in solid-state physics. The simplest example is the Hamiltonian of the Heisenberg model

$$H = \sum_{t,t'} I_{t-t'} (\sigma_t^Z \sigma_{t'}^Z + \sigma_t^x \sigma_{t'}^x + \sigma_t^y \sigma_{t'}^y + 1) + h\,\Sigma(\sigma_t^Z + 1) \tag{20}$$

Here, I_t is a rapidly decreasing function on the lattice Z^ν, σ_t^x, σ_t^y, σ_t^z are Pauli matrices, acting in the exemplar H_t, $t \in Z^\nu$, of two-dimensional complex space, h is a parameter (the external magnetic field), and the operator H itself acts in the tensor product $\underset{t}{\otimes} H_t$ of space H_t with marked vectors $e_{-1}^{(t)}$ ("downwards" spin: the eigenvector of the matrix σ_t^z with eigenvalue -1). It is easily seen that vector $\Omega = \underset{t}{\otimes} e_{-1}$ is the ground state for H, while the vectors $g_{t_0} = \underset{t \neq t_0}{\otimes} e_{-1}^{(t)} \otimes e_{+1}^{(t_0)}$ generate the one-particle invariant subspace H_1 of Hamiltonian H, and the vectors $g_{t_1,t_2} = \underset{t \neq t_1,t_2}{\otimes} e_{-1}^{(t)} \otimes e_{+1}^{(t_1)} \otimes e_{+1}^{(t_2)}$ generate the two-particle invariant subspaces of H. We can similarly obtain any k-particle invariant subspace of H. In each of them, H is a cluster operator.

At the present time, for the case when $I_\tau = 0$ with $|\tau| > 1$ (nearest neighbours), only the spectrum with one-particle subspace has been studied (magnetic waves or magnons), and also the scattering of two magnons and their bound states in the two-particle subspace (see [43] or Akhiezer, Paletminskii, and Bar'yakhtar [44]). Model (20) when $\nu = 1$ and $h = 0$ has been studied quite fully in a series of papers of Babbitt and Thomas [42]. There are various modifications of model (20), which describe the interaction of spin waves with moving particles (say electrons); the corresponding Hamiltonians are also cluster operators. Another class of models, which describe the interaction of an infinite homogeneous medium with rigidly clamped particles (impurities), leads to Hamiltonians which, though formally not cluster (since translational invariance is violated), are similar in their properties to cluster operators.

6.4. Cluster Operators as Such

The concept of a cluster operator, introduced in Section 3, is of general mathematical interest apart from any physical context and admits of various modifications and generalizations (see e.g., Malyshev and Minlos [8]).

First, instead of the group Z^ν (and hence the torus T^ν), we can consider any other commutative locally compact group G along with its dual group G*, e.g., the group R^ν (notice that the corresponding class of cluster operators then contains the n-particle Schrödinger operator). Further, we can weaken estimates (II.3) for the cluster functions, and thus consider the class of cluster operators in the p-representation, not with analytic, but now merely with smooth functions a_γ (infinitely smooth or merely l-smooth). By using this class, we can embrace physical systems with slowly decreasing interaction. Finally, we can consider cluster operators which act, not merely in space L_n with fixed h, but also in "Fok space", i.e., the direct sum $\oplus\, L_n$ of such spaces; this enables systems with a variable number of particles to be considered.

Notice that the direct sums of spaces (objects) of the type

$$H_{(n_i)} = \oplus\, L_{n_i},$$

where n_i is a finite or denumerable set of integers, and the cluster mappings that act between them (morphisms), form a category. This category approach to cluster operators can sometimes prove useful (see Minlos [30]).

We have spoken above about the class of multiplicative cluster operators, to which all the transfer matrices belong. The class forms a semi-group (see Malyshev and Minlos [8]). Another important class of cluster operators (introduced in [8]) is made up of the so-called additive cluster operators (which form a Lie algebra). The cluster expansion of their matrix elements has the form

$$a_{T,T'} = \sum_{\substack{s \subseteq T,\, s' \subseteq T': \\ T \setminus s = T' \setminus s'}} \omega_{s,s'}$$

where $\omega_{s,s'}$ is a function defined on the pairs of subsets (s, s'), s, s' $\subset Z^\nu$, which is invariant with respect to their simultaneous displacements and satisfies the cluster estimate (6.3).

The general hypothesis is that the additive cluster operators form a Lie algebra of the group of invertible multiplicative cluster operators.

To be more precise, given any additive cluster operator H, the one-parameter group exp {tH}, $t \in R^1$, consists of multiplicative cluster operators, and conversely, the generator of any such (bounded) group is an additive cluster operator. This result was originally obtained by Malyshev and Minlos [7], for the case of the transfer-matrix semi-group of the Ising model with continuous time. In a more general form, the result was established by Zoladek [28]. In [27], Zoladek used the result to prove the multiplicative property of the spectrum of the multiplicative cluster operator A: the spectrum $\sigma(A^{\otimes n})$ of any tensor power $A^{\otimes n}$ of the operator is contained in the spectrum of A:

$$\sigma(A^{\otimes n}) \subseteq \sigma(A).$$

This result is similar in its ideas to the analog of the Hunziker-Van Vinter - Zhislin theorem, proved by Zoladek for an additive cluster operator in [17].

SHORT REVIEW OF RECENT INVESTIGATIONS

The general methods of studying a spectral structure of transfer-matrix of Gibbsian fields and related topics have some applications and development in recent papers of authors and their collaborators and students. I give here a short review of these themes. A part of the obtained results is published in the monograph [see 55].

1. The Transfer-matrices of Gibbsian Fields

(1) As a continuation of works [6], [7] transfer-matrices of general Gibbsian fields with compact spin at high temperature are studied in paper [56]. In particular the general condition of existence of k-participle invariant subspace is obtained.

(2) In papers [57], [58] some invariant subspaces of the Hamiltonian of the lattice model of chromodynamics were constructed and studied. The gauge group is SU(N) (N is arbitrary) and isotopic (or "flavor") group is SU(2) or SU(3). Four "leading" bosonic subspaces were considered in [57] for small values of couple constant. These subspaces describe the so-called mesonic states (vector meson, pseudoscalar one in "singlet" and "octet" states with respect to "flavor" symmetry). Moreover dispersions of these particles (dependence of energy on quasimomentum) were calculated in leading terms with respect to couple constant. In work [58] similar considerations were done for baryons and antibaryons subspaces: octet and decaplet states with spin 1/2 and 3/2 correspondingly.

(3) In [59] the two-particle bound states of the transfer-matrix of the Gibbsian field for Ising model with three states were studied (at high temperatures and arbitrary value of "chemical potential").

2. The Random Walk in Random Environment

(1) In the series of papers [60], [62], [63] the Markov model of "random walk in random environment" is studied with the help of investigation of leading branch of spectrum transfer-matrix (stochastic operator) of the whole system, which consists of walking particle (or two particles) and configuration of lattice random field (environment).

The evolution of this Markov system happens so that the particle and the field interact mutually. Here they are got limit theorem for a distribution of position of particle (or two particles) for large time, asymptotics correlations of original field, and also auxiliary so-called field "from particle point of view".

(2) The work [61] is close to this theme: there the random walk of particle interacting with random field is studied. The interaction is given by formula like the Feyman-Kac formula. This model is reduced to the previous one by special cluster expansion and we find similar properties: central limit theorem for position of particle for large time, asymptotics of correlations of field and so on.

3. The Asymptotics of Decay of Correlations for Gibbsian Field at High Temperature

(1) In the works [64], [65], [66] the asymptotics of correlations of form

$$\langle F_A, F_{B+y} \rangle, \quad y \to \infty$$

is considered for Gibbsian field of Ising model (at the temperature). Here F_A, F_B are local functions of field on lattice $\mathbb{Z}^\nu$.The most complete results concern a case when y goes to infinity along some coordinate axes. In work [65] it also considers the case when y goes to infinity along an arbitrary direction. The asymptotics have the usual exponential form

$$C/|y|^\gamma \,.\, \exp\{-(m, y)\}$$

and exponent $\gamma > 0$ and m are calculated.

(2) In the case of ANNI-model and $y = t$ directed along some axes of the lattice is considered in work [67]. Here the asymptotics has oscillating form

$$C/t^{\nu/2} \exp\{-mt\} \sin(at + b)$$

$m > 0$. The results, described above, are obtained by detail investigation of the spectrum of transfer-matrix of the corresponding field, as it was explained in the basic text.

4. Inhomogeneous Random Walk on the Lattice

(1) In papers [68], [69] the model of random walk of particle on the lattice was studied. The transition probabilities for that model almost coincide with ones for homogeneous random walk except for finite neighborhood of origin. The limit theorem for positions of walking particle was obtained. In the case of more than one dimension the distribution of probabilities for the position of particle for large t deviates from gaussian distribution only inside finite area around the origin. In the one-dimensional case another picture arises and a new (non-gaussian) law of distribution appears.

(2) Similar problems are studied in [70] for walk of two interacting particles on three-dimensional lattice.

Both papers are based on the investigation of the spectrum of stochastic operator for each model.

5. Some Models from Solid-state Theory

(1) In work [72] the bound states of two-particle lattice system with weak interaction and strong degenerate law of dispersion of particle are bound. It shows that a number of bound states depends of the degree of degeneracy.

The result is based on calculation of the resolvent of the operator of energy and corresponding Fredholm's determinant.

(2) In [71] a similar problem is solved for three and more quasi-particles with the help of an estimate of quadratic form of the corresponding Hamiltonian.

(3) In contrast with these results in works [81], [82], [83] the case of weak perturbation of non-generate law of dispersion of quasi-particles were investigated and unitary equivalence of perturbed system to the original one was proved – these results are similar to the famous theorem of Iorio-O'Carroll for continuous quantum particles.

(4) In the work [75] the particle weakly interacting with the system of bosons is investigated.

The author studied the behavior of energy of ground state system (polaron) depending on total momentum of system. The results of [75] essentially adds the previous ones, see for example [76].

In [80] the case of two fermions was considered by the same method.

(5) The description of large class of lattice Hamiltonians arising from solid-state theory is contained in work [73]. There are a lot of results concerning such models, examples of very interesting models and many of non-solved problems.

(6) In papers [77], [78], [79] the system consisting of two or three bosons interacting with two-level atoms was studied. All channels of scattering for these systems were constructed and its asymptotic completeness was proved. The described system is an approximation of the famous model "spin-boson" – the system consisting of atom and arbitrary number of bosons. This model describes a phenomena of radiation.

6. A Stochastic Dynamics

The methods described above were applied to stochastic (Glauber) dynamics for finding or spectral decomposition of the generator of such dynamics. Most complete results concerning Glauber dynamics for one-dimensional Ising model are contained in [84]. There all k-particles invariant subspaces of generator were founded and the spectrum of the generator in each of them was described.

References

1. R. A. Minlos and I. G. Sinai, (1970). Study of the spectra of stochastic operators arising in lattice models of a gas, *Teor. Mat. Fizika* **2**, 230–243.
2. V. A. Malyshev, (1978). One particle states and scattering theory for Markov processes. In "Lecture Notes in Math. N 653 Locally interacting systems and their application in biology Proceedings Pushchino, Moscow Region, 1976. (Ed. R. L. Dobrushin, V. L. Kryukov, and A. L. Toom), Springer-Verlag, 173–193.
3. V. A. Malyshev, (1980). Complete cluster expansion for weakly coupled Gibbs random fields. In "Multicomponent Random Systems" *Advances in Probability and related topics* **6**, 505–530.
4. F. H. Abdulla-Zadeh, R. A. Minlos and S. K. Pogosian, (1980). Cluster estimates for Gibbs random fields and some applications. In "Multi-component Random Systems" *Advances in Probability and related topics* **6**, 1–36.

5. V. A. Malyshev and R. A. Minlos, (1979). Some results and problems in the study of infinite particle Hamiltonians. *Colloqia Mathematica societatis Janos Bolyai* **29**, Random Fields, Estergom (Hungary).
6. V. A. Malyshev and R. A. Minlos. Invariant subspaces of clustering operators *I. Journal of Stat. Phys.* 1979 **21**, N 3, 231–242. *II. Commun. Math. Phys. 1981* **82**, 211–226.
7. V. A. Malyshev and R. A. Minlos, (1981). Multiplicative and additive cluster expansion for the evolution of the quantum spin systems in the ground state. *Phys. Lett.* **86A**, 405–406.
8. V. A. Malyshev and R. A. Minlos, (1983). Cluster operators, *Proceedings of the Petrovskii Seminar* **9**, 63–80.
9. S. N. Lakaev and R. A. Minlos, (1979). On the bound states of a cluster operator, *Teor. mat. Fiz.* **30**, 83–93.
10. F. H. Abdulla-Zadeh, (1979). Invariant subspaces of the transfer matrix of a gauge field with group Z_2. In collection, "Interacting Markov processes and their application in biology," Pushchino, 64–73.
11. F. H. Abdulla-Zadeh, (1980). Complete cluster expansion for a lattice Young-Mills gauge field with gauge group U(1), *Teor. mat. Fiz.* **45**, No. 2, 276–279.
12. F. H. Abdulla-Zadeh, (1982). Bound states of transfer matrices of Gibbs fields, in: *Mathematical models of statistical physics, Tyumen*, 67–77.
13. V. A. Malyshev, R. A. Minlos and P. V. Khrapov, The absence of one-particle subspaces in spectrum of transfer-matrix of contour model Trudy 4go Mezhdunarodnogo Simpoziuma po Teorii Information (1984), p 128 (in Russian).
14. I. A. Kashapov and V. A. Malyschev, (1983–84). Cluster expansion and spectrum of the Hamiltonian for lattice fermion models, *Sel. Math. Sov.* **3**, 151–181.
15. H. Zoladek, (1983). Gibbs fields in the nondegenerate case and the spectrum of their stochastic operators, *Candidate Dissertation*. Moscow, MGU.
16. H. Zoladek, (1986). Invariant subspaces of an additive cluster operator, *Dokl. Akad. Nauk SSSR* **287**, No. 1, 41–44.
17. H. Zoladek, (1982). Essential spectrum of an N-particle additive cluster operator, *Teor. mat. fiz.* **53**, No. 2, 216–227.
18. R. A. Minlos and P. V. Khrapov, (1984). Cluster properties and bound states of transfer matrices of the Young–Mills model with compact gauge group, I., *Teor. mat. fiz.* **61**, No. 3, 460–465.
19. P. V. Khrapov. Cluster properties and bound states of transfer matrices of Young–Mills lattice fields, Ms. depos. at VINITI, 13-7-84, No. 5060–84 dep.
20. P. V. Khrapov, (1984). I. Cluster expansion and spectrum of the transfer matrix of the two-dimensional Ising model with a high external field, *Teor. mat. fiz.* **60**, 154–155.

    II. Candidate Dissertation, Moscow, MGU, 1985.
21. S. N. Lakayev, (1986). Some spectral properties of the generalized Friedrichs model, *Proceedings of the Petrovskii Seminar* **II**, 210–238.
22. R. A. Minlos, I. A. Mogilner, (1986). Spectrum analysis and scattering theory for three particle cluster operators. In. *Mathematical problems of stat. mech. and dynamics*. Reidel Publ. corp.
23. Sh. S. Mamatov and R. A. Minlos, (1984). Spectrum of three-particle cluster operators, *Teor. mat. fiz.* **58**, No. 2, 323–328.
24. Sh. Mamatov and R. A. Minlos, (1989). Bound states of two-particle cluster operator. *Theor. i Mat Fiz.*, **79**, No. 2, 163–179, [in Russian].
25. Sh. S. Mamatov, (1984). I, On the bound states of the transfer matrix of the Ising model, *TMF* **2**, 213–218.

    II. The bound states of the transfer matrix of the Ising model with two-step interaction, *Trudy Mosk. Mat. Obshchestva* **49**, 71–94, (1986).

26. I. V. Chernykh, (1985). Some properties of cluster operators, Vestnik MGU, *Ser. I.* **2**, 3–6.
27. H. Zoladek, (1983). Localization of the spectrum of transfer-matrix of Ising model Preprint. Warsaw Univ.
28. H. Zoladek, (1983). On relations between additive and multiplicative clustering operators Preprint. Warsaw Univ.
29. V. A. Malyshev, (1980), Cluster expansions in lattice models of statistical physics and quantum field theory, *Usp. Mat. Nauk.* **35**, 3–35.
30. R. A. Minlos, (1985). Algebra of many particle operators. In "Progress in Physics. V. 10. Statistical Physics and dynamical systems". *Birkhauser,* 1–16.
31. R. A. Minlos and Ya. Ya. Sinai, (1971). Note on the decrease of correlation in a thermodynamic system, Preprint No. 22, MGU.
32. R. A. Minlos, (1986). How our acquaintance with Efim Samolovich Fradkin began and three etudes in honor of his 60th birthday. I. *Quantum Field Theory and quantum Statist.* Essay in honor of the 60th birthday of E. S. Fradkin. London.
33. E. A. Zhizhina. Basic and exited one-particle states of a Young–Mills lattice gauge field. *Theor. i Mat Fiz.*, **74**, No. 2, 218–221, [in Russian].
34. V. A. Malyshev and R. A. Minlos, (1986). The theory of cluster operators, in: *Interacting Markov processes in biology,* Pushchino.
35. V. A. Malyshev and R. A. Minlos, (1985). Gibbs random fields [in Russian], Moscow, Nauka.
36. J. Glimm and A. Jaffe, (1975). Two and Three body equations in quantum field models. *Commun. Math. Phys.* **44**, 293–320.
37. Th. Spenser, (1975). The decay of the Bethe-Salpeter Kernel in quantum field model. *Commun. Math. Phys.* **44**, 143–164.
38. Th. Spencer and F. Zirilli, (1976). Scattering states and bound states in $\lambda P(\varphi)_2$. *Comm. Math. Phys.* **49**, 1–16.
39. J. Dimock and J. P. Eckmann, (1976). On the bound state in weakly couple $\lambda(\varphi^6 - \varphi^4)_2$. *Commun. Math. Phys.* **51**, 41–54.
40. R. Schor, (1984). The particle structure in ν-dimensional Ising models at low temperature. *Commun. Math. Phys.* **59**, 219–233.
II. Glueball spectroscopy in strongly coupled lattice Gauge theory. *Commun. Math. Phys.* **92**, 369–395.
41. O'Carroll, (1984). Convergent expansions for asymptotically degenerate inverse correlation lengths of classical lattice spin-systems. *J. Stat. Phys. V.* **37**, 3–4.
42. D. G. Bablitt and L. E. Thomas, (1977). Ground state representation of the infinite one-dimensional Heisenberg ferromagnet.
II. An explicit Plancherel Formula *Commun. Math. Phys.* **54**, 255–278.
43. M. Reed and B. Simon, (1979). Methods of Modern Mathematical physics V. III, IV; Acad. Press N.Y.
44. A. I. Akhiezer, V. G. Bar'yakhtar and S. V. Peletminskii, (1967). Spin waves [in Russian], Nauka, Moscow.
45. A. M. Polyakov, (1968). Microscopic description of critical phenomena, *Zh. Eksp. Teor. Fiz.* **55**, 1026.
46. Ya. G. Sinai, (1980). Theory of phase transitions [in Russian], Nauka, Moscow.
47. Berezin, (1965). The method of secondary quantization [in Russian], Nauka, Moscow.
48. V. I. Arnol'd, A. N. Varchenko, and S. M. Gussein-Zade, (1984). Singularities of differentiable mappings (II) [in Russian], Nauka, Moscow.
49. L. D. Faddeyev, (1963). Mathematical aspects of the quantum theory of scattering for a three-particle system, *Tr. Mat. in-ta AN SSSR.*

50. S. M. Merkur'ev and L. D. Faddeyev, (1985). Quantum theory of scattering for many-particle systems [in Russian], Nauka, Moscow.
51. Van-Vinter, (1964). Theory of finite systems of particles. *Mat. Fys. Skr. Danske Vid. Selsk.* 1.
52. S. Weinberg, (1964). Systematic solution of multiparticle scattering problems. *Phys. Rev.* **133**, 232–256.
53. R. Durrett, (1981). An introduction to infinite particle systems. *Stochastics processes and their applications,* **11**, 109–150.
54. E. Seiler, (1982). Gauge Theories as a problem of constructive quantum field theory and statistical mechanics. *Lect. Notes in Physics* **159**, Springer-Verlag Berlin, Heidelberg, N-Y.
55. V. A. Malyshev and R. A. Minlos, (1994). Linear infinite-particle operators. *The monograph, Moscow, "Nauka".* There is an English translation, AMS, Providence, 1995.
56. R. A. Minlos, (1991). Invariant k-particles subspaces of Hamiltonian of ciral lattice fields. *Acta Applicandae Mathematicae* **22**, 55–75.
57. R. A. Minlos and E. A. Zhizhina, (1991). Meson states in lattice QCD. *Advances in Soy. Math.* Vol. 5, 113–137. "Many-particle Hamiltonian: spectra and scattering", (Ed. R. Minlos), AMS, Providence.
58. R. A. Minlos and E. A. Zhizhina, (1992). Baryon states in lattice QCD. In *Ideas and methods in quantum and statistical physics.* Vol. 2, 425–430, (Ed. S. Albeverio, J. E. Fenstad, H. Holden and T. Lindstrom), Cambridge, Univ. Press.
59. J. Abdullaev and R. A. Minlos, (1994). An extension of lsing model. *Advances in Soy. Math.* Vol. 20, 1–20, "Probability contributions to statistical mechanics", (Ed. R. Dobrushin), AMS, Providence.
60. C. Boldrighini, R. A. Minlos and A. Pellegrinotti. Central limit theorem for the random walk of one or two particles in random environment with mutual interaction. *ibid.*, 21–76.
61. R. A. Minlos. Random walk of particle interacting with a random field. *ibid.*, 77–90.
62. C Boldrighini, R. A. Minlos, and A. Pellegrinotti, (1994). Interacting random walk on dynamical environment. I. Decay of correlations (one particle). *Annals of Institute of Poincare* Vol. 30, No. 4, 519–558
II. Field "from particle point of view". *ibid.*, 559–605.
III. Decay of correlations (two particles). *Italian Math. Journal* (in press).
63. R. A. Minlos, (1993). Random walk in random environment. In *Proceedings of Conference in Capri,* May 1993. Advances in Dynamical Systems and Quantum Physics, (Ed. S. Albeverio, R. Figari, E. Orlandi, A. Teta), World Scientific, Singapore, N.J, London, Hong Kong, 1995, 194–200.
64. E. A. Zhizhina and R. A. Minlos, (1988). Asymptotics of decay of correlations for Gibbsian spin fields. *Theor. I Mat. Fiz.* **77**, 1, 3, 1–12, [in Russian, there is an English translation].
65. R. A. Minlos and E. A. Zhizhina, (1995). Asymptotics of decay of correlations for lattice spin fields at high temperatures. I. *The Ising model* (submitted to *Journal of Stat. Phys.*).
66. R. A. Minlos, (1993). Spectrum of transfer-matrices of Markov fields and their asymptotic behavior. In *Proceedings of Fall Semester in EMI,* St. Petersburg, (in press).
67. R. A. Minlos and E. A. Zhizhina, (1989). Asymptotics of decay of correlations in the ANNNI model at high temperatures. *Journal of Stat. Phys.* **56**, No. 5/6, 957–963.
68. E. A. Zhizhina and R. A. Minlos, (1994). Local limit theorem for inhomogeneous random walk on the lattice. *Teor. verojt. i ee primen.* **39**, No. 3, 513–529, [in Russian].

69. R. A. Minlos, (1993). Limit theorems for inhomogeneous random walk. In *Proceedings of Spring Semester in EMI*, St. Petersburg.
70. R. A. Minlos and E. A. Zhizhina, (1996). Central limit theorem for inhomogeneous random walks of two particles. *Potential Analysis*, Vol. 5, No. 2, 139–172.
71. A. I. Mogilner, (1988). On weakly bound states of a few quasi-particles in a three-dimensional crystal. *Fiz. nizklch. temperatur.* **14**, 977–980, [in Russian, English translation available].
72. R. A. Minlos and A. I. Mogilner. On bound states of two weakly interacting quasi-particles with strongly degenerate dispersion law. *ibid.*, 1082–1086.
73. A. I. Mogilner, (1991). Hamiltonians in solid-state physics as multiparticle discrete Schrodinger operator: problems and results. *Advances in Soy. Math.* Vol. 5. 139–194, "Many-particle Hamiltonians: spectra and scattering", (Ed. R. A. Minlos), AMS, Providence.
74. R. A. Minlos and A. I. Mogilner, (1992). Some estimates of the spectrum of quasi-particle system. *Journal of Math. Phys.* (in press).
75. R. A. Minlos, (1992). Lower branch of the spectrum of a fermion interacting with a bosonic gas (polaron). *Teor. i Mat. Fiz.*, **92**, 2, 255–268, [in Russian, English translation available].
76. H. Spohn, (1988). The polaron at large total momentum. *J. Phys. A: Math. Gen.* **21**, 1199–1211.
77. R. A. Minlos and H. Spohn (1994). Spectral analysis of spin-boson model. In *Volume of Berezin's Memory*, (Ed. A. Vershik, R. Dobrushin, R. Minlos and M. Shubin), AMS Publishing, Providence (in press).
78. Yu. Zhukov and R. A. Minlos (1994). A system of three bosons interacting with atom. *Theor. i Mat. Fiz.* (in press), [in Russian].
79. Yu. V. Zhukov and R. A. Mintos, (1995). An interaction of system of no more 3 bosons with atom. *Uspechi Mat. Nauk* (in press), [in Russian].
80. A. M Mel'nikov, (1995). On lower branches of the spectrum of a system of two fermions interacting with bosonic gas. *Uspechi Mat. Nauk* (in press), [in Russian].
81. A. M. Mel'nikov and A. I. Mogilner, (1991). Generalization of the Iorio-O'Carroll theorem for the case of lattice Hamiltonians. *J. Phys. A: Math. Gen.* **24**, 3671–3676.
82. Yu. V. Zhukov (1996) The Iorio-O'Carrol theorem for an N-particle lattice Hamiltonian. *Theor. and Math. Phys.* Vol. 107, No. 1, 478–486 [in Russian].
83. Yu. V. Zhukov, (1995). The spectrum of three-particles lattice Hamiltonians. Submitted to *Journal Funct. Analysis i prilozh.* [in Russian].
84. R. A. Minlos and A. Trish, (1995). Complete spectral decomposition of generator of Glauber dynamics for one dimensional Ising model. *Uspechi Mat. Nauk.* 209–210 [in Russian].

Robert Adol'fovich Minlos (1931-2018) – His Work and Legacy

C. Boldrighini (Istituto Nazionale di Alta Matematica, Unità locale Università Roma Tre)
V. A. Malyshev (Faculty of Mechanics and Mathematics, Lomonosov Moscow State University)
A. Pellegrinotti (Dipartimento di Matematica e Fisica, Università Roma Tre)
S. K. Poghosyan (Institute of Mathematics of the NAS RA, Yerevan)
Ya. G. Sinai (Department of Mathematics, Princeton University)
V. A. Zagrebnov (Institut de Mathématiques de Marseille)
E. A. Zhizhina (Institute for Information Transmission Problems, Moscow)

Robert Adol'fovich Minlos (Courtesy of A.Kassian)

On 9 January 2018, the renowned mathematician Professor Robert Adol'fovich Minlos passed away at the age of 86. An eminent researcher and outstanding teacher, he was a world-renowned specialist in the area of functional analysis, probability theory and contemporary mathematical physics.

R. A. Minlos was born on 28 February 1931 into a family associated to the humanities. His father Adol'f Davidovich Miller was known as a lecturer and author of English dictionaries and manuals. His mother Nora Romanovna (Robertovna) Minlos was an historian-ethnographer. Her brother Bruno Robertovich Minlos, with a PhD in historical sciences, was a specialist in the history of Spain. This is perhaps why Robert Adol'fovich loved poetry, wrote verses himself, was a fervent theatre-goer from his school years and was seriously occupied with painting from the age of 40.

Nothing foretold a mathematical future but when he was 15, the young Robert accidentally saw a poster about the Moscow Mathematical Olympiad for schoolchildren. He participated in it, obtained the second prize and, inspired by that, began to attend the school club led by E. B. Dynkin. In 1949, Robert joined the Faculty of Mechanics and Mathematics of the Moscow State University. He continued to participate in Dynkin's seminar, which, together with A. S. Kronrod's seminar, had a great influence on him as an undergraduate student.

R. A. Minlos prepared his first scientific paper (equivalent to a Master's degree thesis) in 1950 while participating in the Moscow State University seminar on the theory of functions of a real variable (under the leadership of A. S. Kronrod). However, the real scientific interests of the young mathematics student began to form after he became acquainted to I. M. Gelfand. Their joint publication 'Solution of the equations of quantum fields' (Doklady Akad. Nauk SSSR, n.s., 97, 209–212, 1954) became Minlos' diploma thesis in mathematics. It was devoted to the *functional* or, in mathematical physics language, the *path* integral, which has a direct relation to quantum physics.

As Minlos himself admitted: "My further life in mathematics was predetermined by that work because I was subsequently mainly occupied with mathematical physics." There were, nevertheless, more works on random processes, on measure theory and on functional analysis. Very soon, one of his papers: 'Extension of a generalised random process to a completely additive measure' (Doklady Akad. Nauk SSSR, 119, 439–442, 1958), brought Minlos to worldwide fame.

It became the basis of his Candidate (equivalent to PhD) Dissertation: "Generalised random processes and their extension to a measure", which was published in Trudy MMO, 8, 497–518, 1959. This result, which is important for the theory of random processes, as well as for functional analysis, is now known as the *Minlos theorem* on the extension of cylindrical measures to Radon measures on the continuous dual of a nuclear space, i.e. the continuation of a process to a measure on spaces adjoint to nuclear spaces.

The connection of Minlos to mathematical physics at that time manifested in the publication (jointly with I. M. Gelfand and Z. Ya. Shapiro) of the monograph "Representations of the rotation and Lorentz groups and their applications" (1958), which was later translated from the Russian by Pergamon, London, in 1964. Note that the monograph appeared in 1958, just at the time when the need for physicists to understand representation theory was strongly motivated by the discovery of elementary particle symmetries, as well as the role of their spins and symmetries related to relativistic Lorentz transformations.

From 1956 to 1992, R. A. Minlos was employed by the Department of the Theory of Functions and Functional Analysis of the Faculty of Mechanics and Mathematics at Moscow State University (MSU). In that period, there was a need to organise a joint seminar with F. A. Berezin, primarily to discuss the mathematical problems of quantum mechanics and quantum field theory.

Seminar of F.A.Berezin and R.A.Minlos at Faculty of Mechanics and Mathematics – MSU, 1959. (Courtesy of N.D.Vvedenskaya)

A real advance of activity in the field of mathematical physics at the Faculty of Mechanics and Mathematics of MSU was achieved with the organisation in 1962 by R. A. Minlos and R. L. Dobrushin of a seminar on statistical physics. It soon became widely known in the Soviet Union and abroad as the *Dobrushin-Malyshev-Minlos-Sinai* seminar. The quantum aspects of statistical mechanics at the seminar were primarily associated to the name of R. A. Minlos. The seminar lasted until 1994 and had a huge impact on the world of modern mathematical physics. Almost all the celebrated specialists in the field visited Moscow during the lifespan of the seminar. Besides traditional scientific contacts with Socialist European countries, a fruitful collaboration was also established with colleagues from other countries. However, the most intensive contacts were within the country, involving almost all the republics. For example, many of Minlos' PhD students came from Uzbekistan.

The beginning of the 1960s was extremely fruitful for Robert Adol'fovich. Initially, there were new results obtained jointly with L. D. Faddeev on the quantum mechanical description of three particles (1961). This was followed by two articles devoted to the study of the thermodynamic limit in classical statistical physics (1967). There, R. A. Minlos suggested the first rigorous mathematical definition of the limiting Gibbs distributions for an infinite system of interacting classical particles and also analysed the properties of such distributions (Funct. Anal. Appl., 1, 140–150 and 206–217, 1967). His approach was very close to the classical Kolmogorov construction of random processes (fields). This result anticipated the origin of the Markovian understanding of Gibbs random fields in the sense of *Dobrushin-Lanford-Ruelle* (1968).

The result (together with Ya. G. Sinai) of the appearance of phase separation in lattice systems at low temperatures (Math. USSR-Sb., 2, 335–395, 1967; Trudy MMO, 17, 213–242, 1967, and 19, 113–178, 1968) was of fundamental importance for the mathematical theory of phase transitions. It formed the basis of Minlos' doctoral dissertation, which he submitted for habilitation in 1968. In another joint work with Ya. G. Sinai (Theor. Math. Phys., 2, 167–176, 1970), the foundation was laid for a new approach to the study of spectral properties of many-particle systems. In combination with cluster expansions, this approach drove significant progress in the description of properties of such infinite systems, including the spectrum of elementary particles of quantum fields and the mathematical description of the *quasi-particle* picture in statistical physics.

The new powerful method of cluster expansions was, from the very beginning, a central one in the list of interests of Robert Adol'fovich. The results of a large series of papers in this topic by R. A. Minlos, V. A. Malyshev and their students have been summarised in two monographs: "Gibbs random fields: cluster expansions" (Springer 1991, translation of the 1985 Russian edition) and "Linear infinite-particle operators" (Amer. Math. Soc. 1995, translation of the 1994 Russian edition). As was outlined in the book "Gibbs Random Fields", the method of cluster expansions provides, besides a construction of the limiting Gibbs measure, a cluster representation of the projections of the limiting Gibbs measure onto bounded regions.

A famous peculiarity of the Dobrushin-Malyshev-Minlos-Sinai seminar was not only its duration of about four hours, which was amazing for foreign guests, or the assertive directness in communicating with lecturers but also the opportunity to obtain from the discussions some interesting problems to be solved. In essence, the seminar was functioning as a *machine*, generating questions and a possible way to convert them into answers. Robert Adol'fovich was always one of the sources of interesting questions and open problems. The list of projects thus originated includes, for example, cluster expansions and their applications to the problem of uniqueness/non-uniqueness of Gibbs states, the quantum three-particle problem, the Trotter product formula for Gibbs semigroups, the study of infinite-particle operators spectra, the analysis of the quasi-particle picture in statistical physics and many others.

The Dobrushin-Malyshev-Minlos-Sinai seminar had a tradition of studying and discussing new important publications on mathematical results and problems in statistical physics. The Trotter product formula problem for Gibbs semigroups was originated after discussing the Pavel Bleher report about new techniques, *reflection positivity* and *infrared bound estimates* (launched by Fröhlich-Simon-Spencer (1976–1978) to prove the existence of phase transitions). In the case of quantum systems, this technique involves the Trotter product formula approximation of the Gibbs density matrix. In this context, Robert Adol'fovich posed a question about the topology of convergence of the Trotter product formula because, to obtain the infrared bound, one has to interchange the trace and the limit of the Trotter approximants. In fact, this operation is not harmful for quantum spin systems since the underlying Hilbert spaces are finite-dimensional but it does produce a problem, for example, in the case of *unbounded* spins. A typical example is the problem of the proof of infrared bounds for the case of structural phase transitions in one-site double-well anharmonic quantum crystals with harmonic interaction between sites (Fröhlich, 1976). Then, the interchange is possible only when the Trotter product formula converges in trace-norm topology. For a particular case of anharmonic quantum crystals, the convergence of the Trotter product formula in the trace-norm topology was proved via the Feynman-Kac representation for Schrödinger (Gibbs) semigroups. The abstract result, which also includes a generalisation to trace-norm Trotter-Kato product formula convergence, is due to H. Neidhardt and V. A. Zagrebnov (1990). So, the answer to the question posed by Robert Adol'fovich was solved affirmatively in favour of trace-norm topology for the case of Gibbs semigroups.

At the end of the 1990s, Robert Adol'fovich returned to the question of quantum phase transitions, in the area of anharmonic quantum crystals (which was already well-known to him) but from the opposite direction. It is known that in contrast to classical systems, phase transitions in their quantum analogues may disappear due to intrinsic quantum fluctuations, which may lead to important tunnelling in the double-well potential. A particular manifestation of that is the elimination by these fluctuations of the order parameter even at zero temperature, whereas it is non-zero in the classical limit when the Planck constant $\hbar = 0$. A typical example is the above structural phase transition in one-site double-well potential anharmonic quantum crystals with harmonic interaction between sites for particles of mass m in each site. Moreover, experimental data for crystals close to this model manifest a so-called "isotopic effect": the order parameter of the structural phase transition for samples with light masses $m < m_c$ disappears, a fact which warms up interest in the mathematical aspect of this phenomenon.

During his visits to Dublin, Leuven and Marseilles, Robert Adol'fovich, in collaboration with E. A. Pechersky, A. Verbeure and V. A. Zagrebnov, addressed the proof of the existence of a critical mass m_c such that below this threshold the quantum state of the system is in a certain sense *trivial*, or at least the order parameter is trivial. In two papers, with A. Verbeure and V. A. Zagrebnov (2000) and then with

R.A.Minlos with co-authors N.Angelescu, and V.A.Zagrebnov.
Visit to Dublin Institute for Advanced Studies, 2000.

E. A. Pechersky and V. A. Zagrebnov (2002), R. A. Minlos proposed using cluster expansion techniques for the small parameter $\xi = \sqrt{m}/\hbar$. Then, the classical limit corresponds to $\xi \to \infty$ and the quantum regime, with zero order parameter for any temperature, corresponds to $\xi < \sqrt{m_c}/\hbar$. Since the structural phase transition in the model manifests as the displacement order parameter, these papers consider projection of the full quantum state on the commutative coordinate $*$-subalgebra $\mathfrak{A}_q$ of bounded functions of displacements on the lattice. Then, the Feynman-Kac-Nelson formula for the Gibbs semigroup kernel allows one to show that this projection reduces to a classical ensemble of weakly interacting (for $m < m_c$) Ornstein-Uhlenbeck trajectories. Using the cluster expansion technique, the exponential mixing of the limit state with respect to the lattice group translations was proven for all temperatures, including zero, if $m < m_c$ (2000). To check that in this domain of light masses (high *quantumness*) the order parameter is zero for all temperatures, including the ground state, R. A. Minlos and his coauthors, in the 2002 paper, used the external sources h conjugated to the local displacements instead of localising the trajectories boundary conditions, as in the 2000 paper. This allows the analysis, for any temperature θ, of the free-energy density function $f(\theta, h)$ for free or periodic boundary conditions. It is proved in the 2002 paper that there exists a radius $h_0(m)$ such that $h \mapsto f(\theta, h)$ is holomorphic in the disc $\{h \in \mathbb{C} : |h| < h_0(m)\}$ for any $\theta \geq 0$ if $m < m_c$. Moreover, the Gibbs expectations: $h \mapsto \langle A \rangle(\theta, h)$, are holomorphic in the same disc for any bounded operator A of a quasi-local $*$-algebra $\mathfrak{A}$ of observables. The analyticity, in particular, yields that the displacement order parameter is equal to zero for $h = 0$ and for any temperature $\theta \geq 0$ if $m < m_c$.

This was also a period when, during his visits to KU Leuven and CPT Marseilles, Robert Adol'fovich got into an argument with A. Verbeure about the mathematical sense of the notion of *quasi-particles* in many-body problems and of the *corpuscular* structure of infinite system excitations. In the framework of quantum statistical mechanics, an attractive way to promote this notion was based on the non-commutative central limit theorem for *collective* excitations. This concept yields a plausible (for physics) picture of *boson* quasi-particles excitations (phonons, magnons, plasmons, etc.) in the corresponding Fock spaces (A. Verbeure et al. (1995)).

On the other hand, in their book *Linear infinite-particle operators*, V. A. Malyshev and R. A. Minlos proposed the description of a quasi-particle picture based on the construction by cluster expansions of the lower branches of the spectrum of infinite many-body systems with good clustering. This idea goes back to the paper by R. A. Minlos and Ya. G. Sinai "Investigation of the spectra of some stochastic operators arising in the lattice gas models" (1970). There, a new approach to studying the spectral properties of the transfer matrix in general lattice models at high temperatures was developed. For translation-invariant systems, the lowest branch of the spectrum enumerated by momentum corresponds to one-quasi-particle excitations above the ground state. Then, in the simplest case, the energy of these excited states is completely defined by the momentum. This is called a dispersion law for quasi-particles, which is also well-known for boson quasi-particles. If the system possesses a good clustering, one can construct separated translation-invariant two-, three- and more (interacting) quasi-particles excited states, which are combinations of branches with bands of continuum spectra. Robert Adol'fovich called this property of excitations "The 'corpuscular' structure of the spectra of operators describing large systems" (the title of his paper in *Mathematical Physics* 2002, Imperial Coll. Press, 2000).

The technique developed by V. A. Malyshev and R. A. Minlos allowed the study of the *corpuscular* structure of generators of stochastic dynamics. Their approach was also applied to generators of stochastic systems: Glauber dynamics, the stochastic models of planar rotators, the stochastic Ising model with random interaction and other lattice stochastic models with compact and non-compact spin spaces, as well as stochastic dynamics of a continuous gas and other stochastic particle systems in the continuum. Using this technique, one can find spectral gaps and construct lowest one-particle invariant subspaces of the generator that determine the rate of convergence to the equilibrium Gibbs state. Moreover, it also allowed the study, in detail, of the spectrum branches of infinite-particle operators on the leading invariant subspaces and, in particular, the construction of two-particle bound states of the cluster operators. These results led to the understanding that a wide class of linear infinite-particle operators of systems in a regular regime have a corpuscular structure.

Visiting Leuven and Marseilles, Robert Adol'fovich proposed elucidating the concept of corpuscular structure of spectral branches for several particular models on the lowest level of one-particle elementary excitations. This programme was performed in papers with N. Angelescu and V. A. Zagrebnov (2000, 2005) and then all together with J. Ruiz (2008), for lattice models and for polaron-type problems. More activity and results in this direction were due to intensive collaboration of Robert Adol'fovich with the group of H. Spohn, where he studied spectral properties of Hamiltonians for quantum physical systems, in particular for Nelson's model of a quantum particle coupled to a massless scalar field.

Another long-term and fruitful collaboration of R. A. Minlos was with the Bielefeld group, essentially with Yu. G. Kondratiev and his co-authors and pupils. Firstly, they generalised the original method (adapted for lattices) to functional spaces to control general particle configurations. This allowed the extension of their analysis from lattice to *continuous* systems. In the paper "One-particle subspace of the Glauber dynamics generator for continuous particle systems" (2004), they studied, in detail, the spectrum of the generator of Glauber dynamics for continuous gas with repulsive pair potential. To this end, the invariant subspaces corresponding to the corpuscular structure were constructed.

The ideas and technical tools elaborated in this paper were used in a number of other projects (Yu. Kondratiev, E. Zhizhina, S. Pirogov and O. Kutoviy) on equilibrium and non-equilibrium continuous stochastic particle systems. This, in particular, concerns a delicate continuous models problem of thermodynamic limit. One unexpected application of this technique concerns image restoration processing. Robert Adol-fovich, in collaboration with X. Descombes and E. Zhizhina, actively participated in the INRIA project on the mathematical justification of a new stochastic algorithm for object detection problems. The result was summarised in the article "Object extraction using stochastic birth-and-death dynamics in continuum" (2009).

In addition to the Dobrushin-Malyshev-Minlos-Sinai seminar in the 1970s, there was also a regular tutorial seminar, which was led by Robert Adol'fovich once a week. This was a very good opportunity to learn elements of topological vector spaces, in particular the Minlos theorem about the extension of a generalised random process to a measure on spaces adjoint to nuclear spaces. The seminar also covered elements of mathematical statistical physics in the spirit of the famous "Lectures on statistical physics" in Uspekhi Math. Nauk (1968). These lectures of Robert Adol'fovich very quickly became a textbook for many students and scientists interested in mathematical statistical physics. In these lectures, Robert Adol'fovich systematically used the notion of configuration space, which appeared in his earlier work, where he gave the mathematical definition of the limiting Gibbs measure as a measure on the space of locally finite configurations in $\mathbb{R}^d$. This concept is technically very useful and is close to modern random point process theory.

R. A. Minlos, Ya. G. Sinai and R. L. Dobrushin were often invited by the Yerevan State University and the Institute of Mathematics of the Armenian Academy of Sciences to give lecture courses on statistical mechanics. They all had PhD students working at the Institute of Mathematics in Yerevan. This was the main motivation for the Institute of Mathematics to organise regular conferences (every 2–3 years) in Armenia under the name "Probabilistic methods in modern statistical physics". The first one was held in 1982 and the last one in 1988, three years before the collapse of the Soviet Union.

The conferences restarted in 1995 at the international level. Robert Adol'fovich participated (as a rule, with his students) in all of them, including the conference in Lake Sevan in 2006. He always supported

R.A.Minlos with participants of the conference "Probabilistic methods in modern statistical physics", Yerevan – Lake Sevan, 2006.

the conferences in Armenia by being a permanent member of the programme committee and one of the main speakers, formulating new problems and generating interesting ideas, questions and discussions. Unfortunately, he was not able to participate at the conferences after 2006.

R.A.Minlos with participants of the conference "Probabilistic methods in modern statistical physics": S.Lakaev, V.Zagrebnov, H.Suqiasian, and B.Nakhapetian, Yerevan – Lake Sevan, 2006.

In the early 1990s, Robert Adol'fovich began his collaboration with Italian institutions and mathematicians. He was a guest of the Department of Mathematics at the University of Rome "La Sapienza" many times and he also visited other institutions in Trieste, Naples, L'Aquila and Camerino. During his stay at "La Sapienza", he read a course on the mathematical foundations of statistical mechanics, which was published as a book by the American Mathematical Society in 2000 under the title "Introduction to mathematical statistical physics". In Rome, he began a collaboration with C. Boldrighini and A. Pellegrinotti on models of random walks (RW) in interaction with a random environment fluctuating in time ("dynamic environment").

At that time, several important results on random walks in a fixed environment had already been obtained, by Solomon, Kesten, Sinai and others, but very little was known for dynamic environments. Following the usual terminology, the behaviour of RW for a fixed choice of the history of the environment is called "quenched" and its distribution induced by the probability measure of the environment is called "annealed". A first result had been obtained by C. Boldrighini, I. A. Ignatyuk, V. A. Malyshev and A. Pellegrinotti on the annealed model of a discrete-time random walk on a d-dimensional lattice in mutual interaction with a dynamic random environment. Robert Adol'fovich proposed applying the results that he had obtained, together with V. A. Malyshev and their students, on the spectral analysis of the transfer matrix for perturbed homogeneous random fields. The approach proved to be very fruitful and in the following years 1993–1996, several results (C. Boldrighini et al. (1994)) were obtained on the annealed RW, on the convergence to a limiting measure for the field "as seen from the particle", on the decay of the space-time correlation for the random field in interaction with the RW and on the RW of two particles in mutual interaction with a random environment.

It was then possible, with the help of some tools of complex analysis of which Robert Adol'fovich had a deep knowledge, to deal with the quenched model of the RW. After the first results of a perturbative approach (C. Boldrighini et al. (1997)), a complete non-perturbative answer could be obtained for the case when the components of the dynamic environment $\xi = \{\xi(x, t) : (x, t) \in \mathbb{Z}^d \times \mathbb{Z}\}$ are independent, identically distributed random variables (C. Boldrighini et al. (2004)). Unlike the case with fixed environment, the quenched RW in dynamic environment behaves almost surely as the annealed RW in all dimensions $d \geq 1$. In low dimension $d = 1, 2$, the random correction to the leading term of the RW asymptotics is of an "anomalous" large size. A quenched local limit theorem was also obtained, with an explicit dependence on the field as seen from the particle. The results were then extended to models of directed polymers in dimension $d > 2$ below the stochastic threshold. Results for a quenched model of RW in a dynamic environment with Markov evolution were also obtained in dimension $d > 2$ by cluster expansion methods (C. Boldrighini et al. (2000)). Further results on models of RW in a dynamic environment were obtained by Zeitouni, Rassoul-Agha, Liverani, Dolgopyat and others.

Later on (in collaboration with F. R. Nardi), it was possible to derive Ornstein-Zernike asymptotics for the correlations of a Markov field in interaction with a RW (C. Boldrighini et al. (2008)) and also for a general "two-particle" lattice operator (C. Boldrighini et al. (2011)).

In the last few years, the interest of Robert Adol'fovich in the study of random walks in a dynamic random environment did not fade and several difficult problems were solved. They concern extensions to

R.A.Minlos with A.Pellegrinotti and C.Boldrighini, Yerevan, 2006.

continuous space (C. Boldrighini et al. (2009)), to continuous time and to the case of long-range space correlations for the environment (in collaboration with E. A. Zhizhina).

Robert Adol'fovich was a wonderful teacher and a patient and wise mentor. Directness, accessibility and enthusiasm attracted numerous students and followers to him. Many of his later PhD students made their first acquaintance with special branches of mathematics and mathematical physics due to the tutorial seminar at the Faculty of Mechanics and Mathematics at MSU. There, they benefited from direct generous contact with the *Master*. This student seminar was combined with lectures and scientific seminars guided by Robert Adol'fovich, together with F. A. Berezin and then with V. A. Malyshev. The lecture notes gave rise to many nice and popular tutorial books, for example "Introduction to mathematical statistical physics", published by R. A. Minlos in Univ. Lect. Series, Vol. 19, AMS 2000. Many of Minlos' former students successfully continue research in different branches of mathematics and mathematical physics, for example: S. K. Poghosyan and E. A. Zhizhina – spectral theory of infinite systems and mathematical problems of statistical mechanics; S. Lakaev – operator spectral theory and mathematical quantum mechanics; A. Mogilner – mathematical biology; E. Lakshtanov – infinite particle systems; and D. A. Yarotsky – random processes and spectral theory of infinite systems.

Besides the students and the tutorial seminar, Robert Adol'fovich was in contact with followers and co-authors assisting the crowded Dobrushin-Malyshev-Minlos-Sinai research seminar. There, Minlos initiated a number of projects, often related to discussions during the seminar. Always attentive and gentle, Robert Adol'fovich shared his enthusiasm to encourage followers in solving the problems.

In this way, Minlos launched the project "Cluster expansions" with V. A. Malyshev. In fact, this happened by accident when they were both in a lift while attending a seminar in the tall main MSU building. Less unusual were the origins of the projects "On the spectral analysis of stochastic dynamics" with E. A. Zhizhina and "Gibbs semigroups" with V. A. Zagrebnov, which in fact started from questions formulated during and after the seminar. The origins of many of them were due to active contacts made by Robert Adol'fovich travelling to other research centres. This is, for example, the case for the project "Application of the spectral analysis of the stochastic operator to random walks in dynamic random environments" with C. Boldrighini and A. Pellegrinotti and "Spectral properties of multi-particle models" with H. Spohn, as well as "Infinite dimensional analysis" and "Stochastic evolutions in continuum" with Yu. G. Kondratiev.

To his students and collaborators, Robert Adol'fovich was a *Master*, who, like a brilliant sculptor, could create a mathematical masterpiece from a shapeless block by cutting off the excess. Sometimes, it brought not just a feeling of amazement but a sense of miracle when, as a result of some incredible expansions, evaluations, virtuoso combinations with various spaces and other technical refinements, complex infinite-dimensional and infinite-particle systems took an elegant, precise and easily understandable form.

In this connection, problems related to the theory of operators and to quantum physics should be especially noted. This theme began in his joint paper with I. M. Gelfand and, since then, it has permanently been the focus of Minlos' attention. In 2010, together with his old co-author and friend V. A. Malyshev, he turned to a fundamental question in quantum chemistry: what is the interaction between atoms? (Theoretical and Mathematical Physics, 162, 317–331, 2010.) However, his favourite subject since the 1960s and until recently has been the quantum three-body problem and point interaction. A long paper ("A system of three quantum particles with point-like interactions", Russian Math. Surveys, 69, 539–564, 2014) was published by R. A. Minlos on this topic.

A recent paper by Robert Adol'fovich was dedicated to another of his favourite subjects: the random walk in a random environment ("Random walk in dynamic random environment with long-range space correlations", Mosc. Math. J., 16:4 (2016), 621–640, with C. Boldrighini and A. Pellegrinotti). His very last manuscript (with C. Boldrighini, A. Pellegrinotti and E. A. Zhizhina) was also on this subject: "Regular and singular continuous time random walk in dynamic random environment".

Robert Adol'fovich selflessly served science and, in everyday life, was a generous and friendly person. He gladly shared his enthusiasm and energy with his students and colleagues. In addition to the accuracy of reasoning and complicated techniques involved, there is always a beautiful idea and harmony in his works. It is interesting to mention his response to the question of Natasha Kondratyeva: "What three mathematical formulas are the most beautiful to you?" He gave the

R.A.Minlos, Moscow, December 2016 (Courtesy of E.Gourko)

answer: "The Gibbs formula, the Feynman-Kac formula and the Stirling formula." And those are the formulae that were widely used by Robert Adol'fovich in his works.

Robert Adol'fovich was notable for his figurative Russian language and good wit, often with subtle mathematical humour. In the 1980s, in a conversation with Roland Dobrushin at the Fourth Vilnius Conferences on Probability Theory and Mathematical Statistics (1985), he expressed his doubt that "the life of a Soviet citizen is *complete* with respect to the *norm* of the anti-alcohol campaign". A campaign was ongoing at that time in the country under the slogan "Sobriety is the norm of our life!" and was visible everywhere on white-red streamers. Since then, this allusion to the completeness of life and normed spaces has entered into the folklore of the mathematical community.

Always surrounded by relatives and intimates, and also by loving pupils, colleagues and friends, Robert Adol'fovich Minlos lived a complete life. In each of those who knew Robert Adol'fovich, he left a bright drop of memory of himself.

www.ingramcontent.com/pod-product-compliance
Lightning Source LLC
Chambersburg PA
CBHW061839220326
41599CB00027B/5343
* 9 7 8 1 9 0 4 8 6 8 9 9 6 *